RIGGING HANDBOOK

The complete illustrated field reference

Jerry A. Klinke

THIRD EDITION
1st printing: May 2007
2nd printing: April 2008
3rd printing: December 2008

Published and distributed by:

ACRA Enterprises, Inc.
2769 West Glenlord
Stevensville, Michigan 49127
800-992-0689
(269) 429-6240

www.acratech.com

ISBN 978-1-8-8872402-8

Rigging Handbook

The complete illustrated field reference

ISBN 978-1-8-8872402-8

Third Edition - 3nd Printing

Release date: May 2007

see page 231 for corrections made since the first printing

This publication IS NOT intended to replace proper, adequate training related to crane and rigging operations. **This publication DOES NOT provide a comprehensive or exhaustive list of all the possible situations encountered with rigging and hoisting operations.** This publication is intended to assist the user by providing typical equipment capacities and general guidelines for job planning. **Always select and use rigging equipment by the rated capacity shown on the ID tag, equipment labels or as provided by the manufacturer.**

Published by:

ACRA Enterprises, Inc.

"Rigging Training and Publications"

2769 West Glenlord, Stevensville, Michigan 49127

800-992-0689 or 269-429-6240

www.acratech.com

DISCLAIMER:

The information contained in this publication was compiled from sources believed to be reliable. It should not be assumed that this material covers all regulations or standards. The author and publisher cannot guarantee correctness or completeness and accepts no responsibility in the use or misuse of this information.

Introduction - by Jerry Klinke

The RIGGING HANDBOOK is a clear, illustrated reference source for rigging professionals, crane operators, and others that perform rigging and hoisting operations. This handbook provides concise, simple answers to rigging situations that may otherwise appear complex in nature. Both apprentices and journeymen will appreciate the simple layout, organization and detailed illustrations provided in this book.

This newest edition includes the latest industry information on rigging and hoisting equipment and techniques. You will notice a heavy emphasis on the ASME B30 standards that cover cranes, rigging and related topics throughout this book. The current OSHA standard that addresses rigging (29 CFR 1910.184) is nearly 30 years old and does not deal with current equipment or industry practices. ASME clearly represents the current leader of safety practices of the industry, and updates the B30.9 standards on a regular basis.

This book tries to cover many of the "problem" areas related to poor rigging practices, based partially from lessons learned through recent industry events, accident investigations, and discussions with other industry experts. **However, the information contained in this book is not a substitute for proper training and education. The proper blend of classroom and hands-on training with the right mix of on-the-job training will help develop competent and skilled rigging professionals.**

I am a firm believer that the better we communicate lessons learned, unique ideas and good rigging practices to others, the safer the rigging industry will become. As a member of the Association of Crane and Rigging Professionals (www.acrp.net) I have had the opportunity to meet with and learn a great deal from many of the other professionals involved in safety and training within the rigging and lifting industry. I invite anyone in the crane and rigging industry to join ACRP and help us to improve the crane and rigging culture of the 21st century.

Contents

SLINGS

Disclaimer

The information contained in this publication was obtained from sources believed to be reliable at the time this publication was written. It should not be assumed that this material covers all the regulations, or standards used in the industry. Suggested procedures should not, therefore, be used without first securing competent engineering advice for any given application.

The publisher and author make no representation or guarantee as to the correctness or sufficiency of any information contained herein, nor a guarantee of results based upon the use of this information, and disclaims all warranties whether implied, express or statutory, including without limitation, implied warranties of merchantability, fitness for use and fitness for a particular purpose.

You assume the entire risk as to the use of this information, and the publisher and the author assumes no liability in connection with either the information presented or use of the suggestions made in this publication.

Basic Hitches

Graphics ©2007 Jerry Klinke

The illustration below shows the 3 basic sling hitches.

VERTICAL

Wire rope / Web Sling

eye & eye

Chain Sling

Web / Roundsling

endless

CHOKER

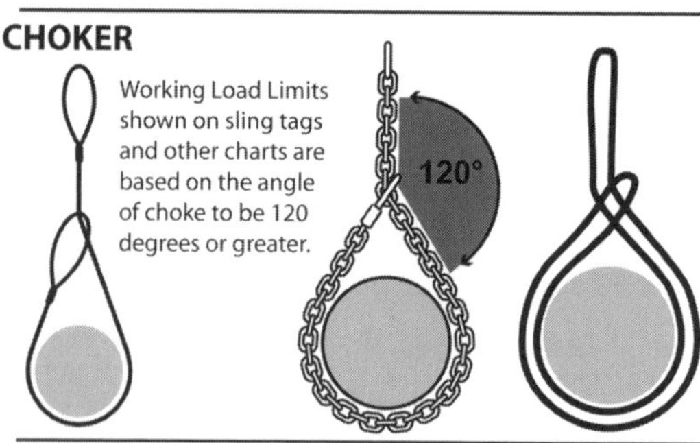
Working Load Limits shown on sling tags and other charts are based on the angle of choke to be 120 degrees or greater.
120°

BASKET

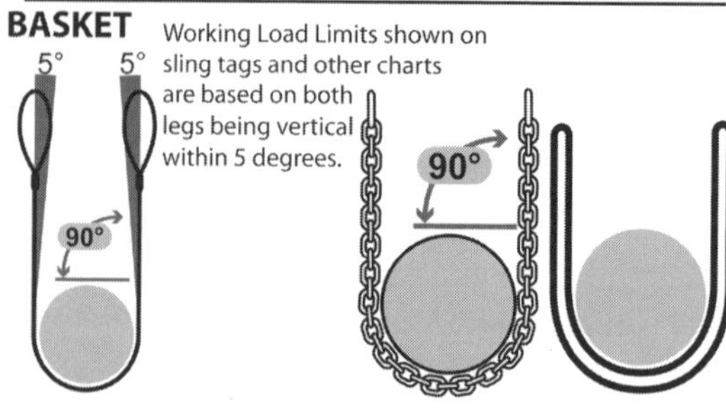
5°
5°
Working Load Limits shown on sling tags and other charts are based on both legs being vertical within 5 degrees.
90°
90°

Hitch Reductions

Capacity charts and the sling tags show the Working Load Limit (WLL) for each hitch based on the examples in the left column.

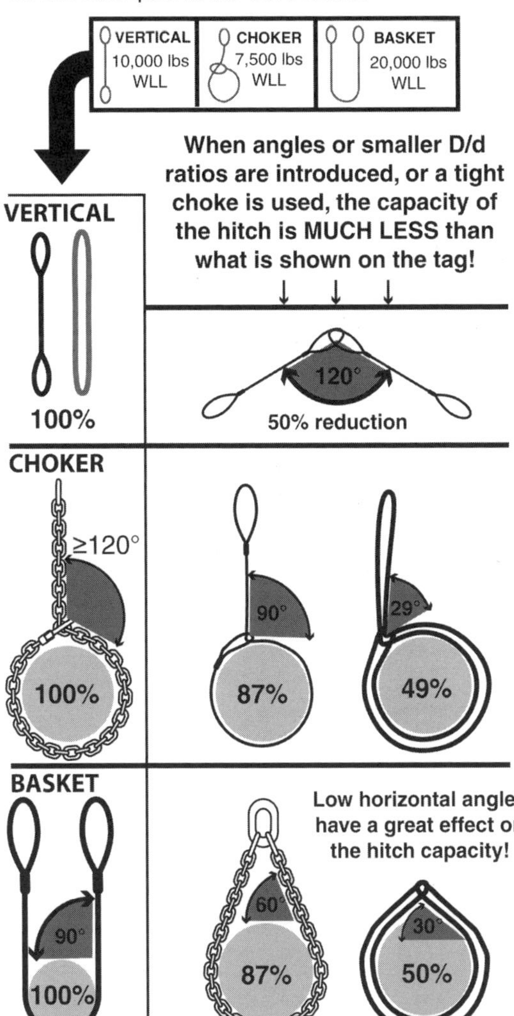
VERTICAL
10,000 lbs
WLL
CHOKER
7,500 lbs
WLL
BASKET
20,000 lbs
WLL
When angles or smaller D/d ratios are introduced, or a tight choke is used, the capacity of the hitch is MUCH LESS than what is shown on the tag!
VERTICAL
100%
120°
50% reduction
CHOKER
≥120°
100%
90°
87%
29°
49%
BASKET
90°
100%
Low horizontal angles have a great effect on the hitch capacity!
60°
87%
30°
50%

All Working Load Limits (WLL) shown are based upon the items being in new, unused condition.

All rigging equipment is subject to wear, misuse, and overloading that may reduce the rated capacity of the equipment. It is required by OSHA and ASME standards that rigging equipment be regularly inspected to determine its condition. A competent person or engineering professional should then determine if the equipment can continue to be used at the rated capacity.

WLL's are based on the following factors:

(1) material strength
(2) design factor
(3) type of hitch
(4) angle of loading
(5) diameter of curvature (D/d)
(6) fabrication efficiency

ASME B30.9-1.7.1 (2003) Identification Requirements

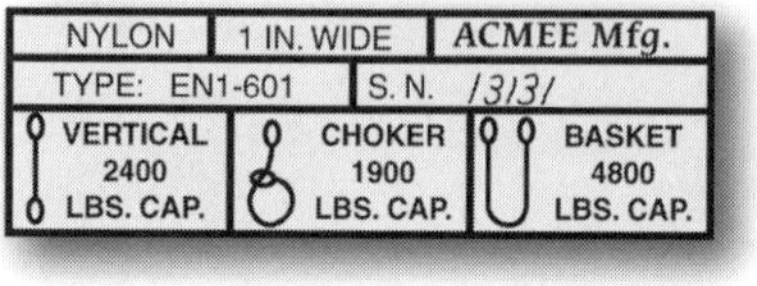

NYLON	1 IN. WIDE	ACMEE Mfg.
TYPE: EN1-601	S. N. 13131	
VERTICAL 2400 LBS. CAP.	CHOKER 1900 LBS. CAP.	BASKET 4800 LBS. CAP.

ALL slings shall be marked to show:
(1) name or trademark of manufacturer
(2) manufacturer's code or stock number
(3) rated loads for the type(s) of hitch(es) used and the angle upon which it is based

ALWAYS read the TAG!

Additional marking requirements for . . .
Wire Rope: diameter or size
Web Slings: synthetic web material used
Roundslings: core and cover material used
Chains: grade, chain size, number of legs, length (reach)

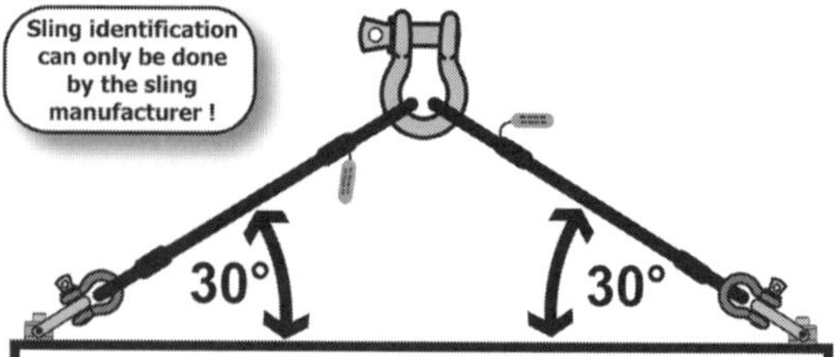

Sling identification can only be done by the sling manufacturer !

30° 30°

ASME B30.9-1.5.3 states "Horizontal sling angles less than 30 degrees SHALL NOT be used except as recommended by the sling manufacturer or a qualified person."

D/d ratio with wire rope slings used in a basket hitch

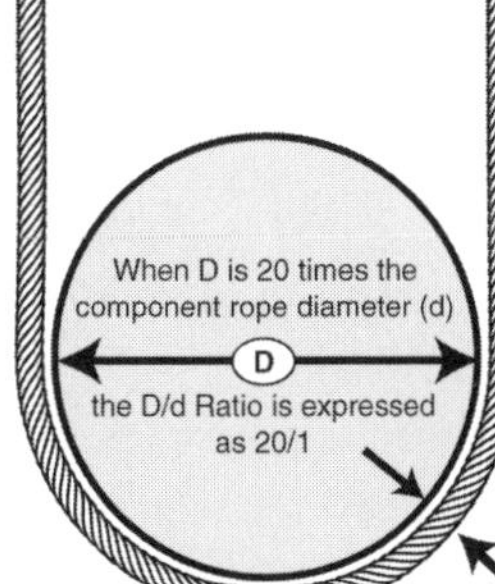
When D is 20 times the component rope diameter (d)
D
the D/d Ratio is expressed as 20/1
d

There is a great misconception about what is a "true" basket hitch. A true basket hitch is one in which a wire rope sling is in contact with a round surface that has a diameter of 20 or more times the rope diameter (assuming the wire rope requires a 20:1 D/d)

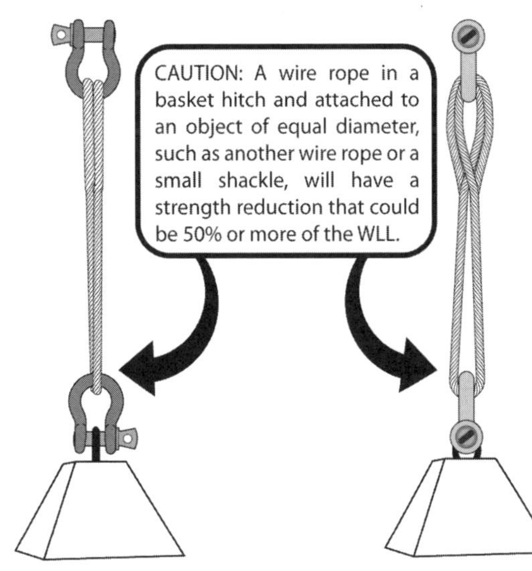
CAUTION: A wire rope in a basket hitch and attached to an object of equal diameter, such as another wire rope or a small shackle, will have a strength reduction that could be 50% or more of the WLL.

NOTE: A sling wrapped around a rectangular surface, is NOT a true basket and will have an effect on the capacity of the sling.

Basket Hitch D/d

The values listed for wire rope slings used in a basket hitch are generally based on a minimum D/d ratio of 20:1.

D = The diameter of the object in load contact with the sling

d = The diameter of the wire rope.

The example on the right shows a "TRUE" basket hitch with a 20:1 D/d ratio and the legs are within 5 degrees from vertical.

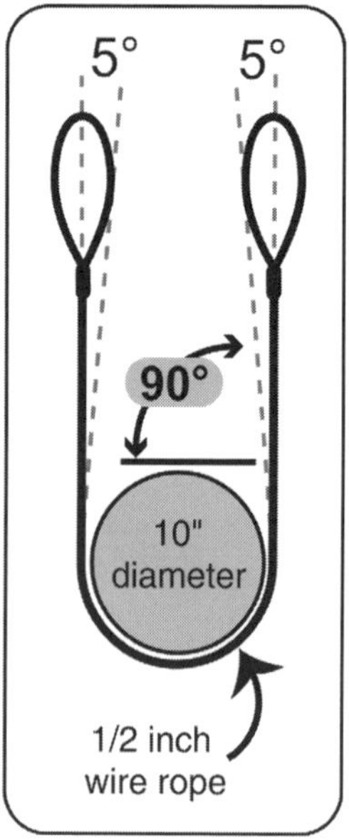

D/d RatioStrength Efficiencies	
D/d	Efficiency
30	95%
20	92%
10	86%
5	75%
2	65%
1	50%

5°
5°
90°
10" diameter
1/2 inch wire rope

To calculate strength efficiencies:

RC = B x E

B: Basket Rating
E: Efficiency (from table)
RC: Reduced basket rating

Choker Hitch Reductions

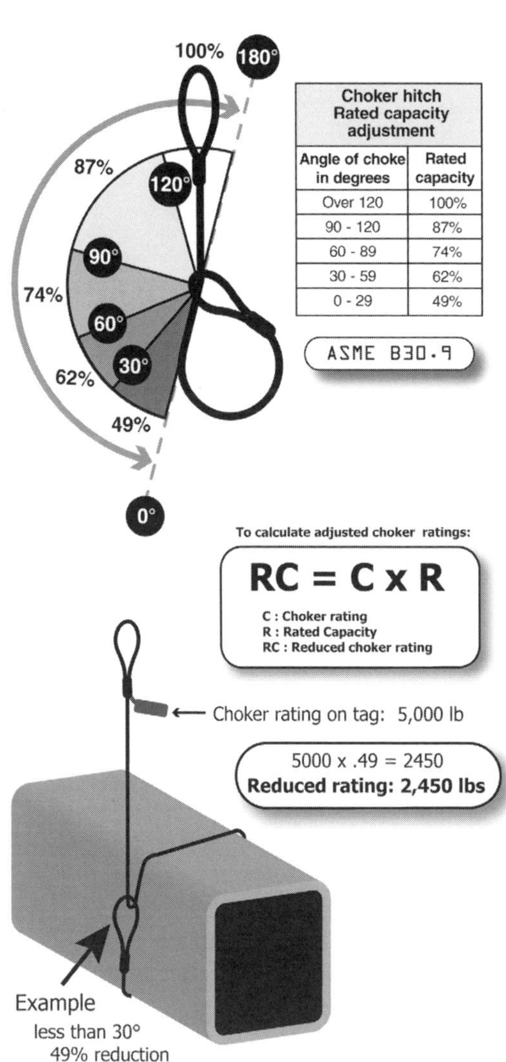
100%
180°
87%
120°
90°
74%
60°
30°
62%
49%
0°
ASME B30.9
To calculate adjusted choker ratings:
RC = C x R
C : Choker rating
R : Rated Capacity
RC : Reduced choker rating
Choker rating on tag: 5,000 lb
5000 x .49 = 2450
Reduced rating: 2,450 lbs
Example
less than 30°
49% reduction

Choker hitch Rated capacity adjustment	
Angle of choke in degrees	Rated capacity
Over 120	100%
90 - 120	87%
60 - 89	74%
30 - 59	62%
0 - 29	49%

Choker Reduction Table

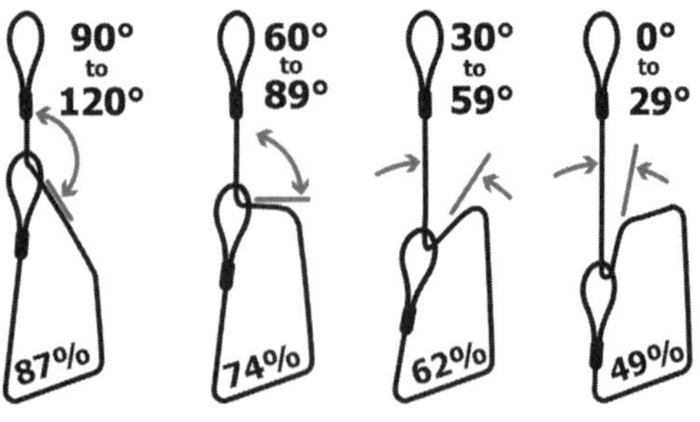
90° to 120°
60° to 89°
30° to 59°
0° to 29°
87%
74%
62%
49%

Choker WLL	Universal Reduction Table			
(shown on tag)	Reduced Capacities by percentage			
	87%	**74%**	**62%**	**49%**
500	435	370	310	245
1,000	870	740	620	490
2,000	1,740	1,480	1,240	980
2,500	2,175	1,850	1,550	1,225
3,000	2,610	2,220	1,860	1,470
3,500	3,045	2,590	2,170	1,715
4,000	3,480	2,960	2,480	1,960
4,500	3,915	3,330	2,790	2,205
5,000	4,350	3,700	3,100	2,450
5,500	4,785	4,070	3,410	2,695
6,000	5,220	4,440	3,720	2,940
6,500	5,655	4,810	4,030	3,185
7,000	6,090	5,180	4,340	3,430
7,500	6,525	5,550	4,650	3,675
8,000	6,960	5,920	4,960	3,920
8,500	7,395	6,290	5,270	4,165
9,000	7,830	6,660	5,580	4,410
9,500	8,265	7,030	5,890	4,655
10,000	8,700	7,400	6,200	4,900

Here is a quick table to help determine the reduced capacity of a choker hitch. Find the closest WLL for the sling you are using (use a lower value if your exact WLL is not shown) and then find the reduced rating under the appropriate angle column.

The Capacity tables shown in this book are only intended to assist the user by providing typical sling capacities for pre-job planning activities.

Be aware than the sling capacities are different depending on each sling manufacturer!

For example; the capacity for a 1" 6x19 wire rope sling can range from 6.7 tons to over 11 tons, depending on the grade and manufacturer.

Never assume that all slings are the same!

This is why it is VERY IMPORTANT to always select and use all slings and rigging hardware by the rated capacity (WLL) shown on the ID tag as provided by the manufacturer.

CAUTIONS & WARNINGS

All ratings shown are based upon the items being new or in "as new" condition. All rigging equipment is subject to wear, misuse, overloading, corrosion, deformation, or alterations that may require a reduction in the rated capacity of the equipment. It is recommended, and required by OSHA and ASME Standards, that all rigging equipment be regularly inspected by a competent person to determine its condition.

WIRE ROPE SLINGS

Wire Rope Grades

IPS { Improved Plow Steel: lowest minimum breaking force

EIPS { Extra Improved Plow Steel: minimum breaking force typically 15% higher than IPS

EEIPS { Extra Extra Improved Plow Steel: minimum breaking force typically 10% higher than EIPS

Wire Rope Core

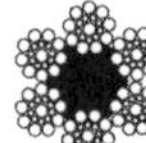

FIBER
A core composed of synthetic fibers .

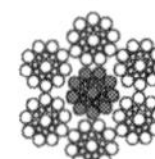

IWRC
A steel core, usually another wire strand.

Wire Rope Splice (connection)

HT Hand Tucked Splice: a loop or eye formed in the end of a rope by tucking the end of the strands back into the main body of the rope in a prescribed manner.

MS Mechanical Splice: swaging one or more metal sleeves over the wire rope to form a loop or eye.
Note: A Femish Eye is a mechanical splice, it is formed by opening up the rope end and reforming it to create a loop or eye then a metal sleeve is slipped over the ends of the splice and mechanically compressed.

The length of a standard wire rope slings is taken from bearing point to bearing point.

Length of sling

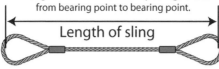

Gromment slings, also called endless slings, are usually manufactured and shown on engineering drawings using a measurment that is the circumference of the sling and not the bearing point distance.

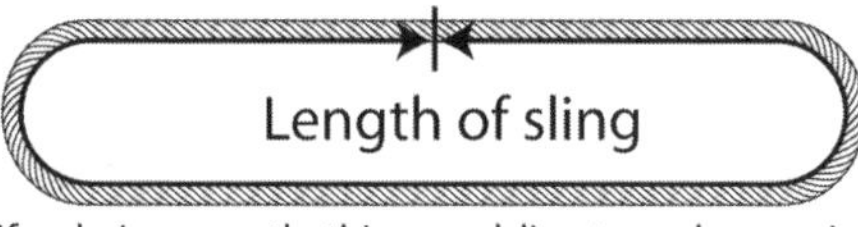
Length of sling

If ordering a synthethic roundsling to replace a wire rope gromment sling, you need to inform the sling manufacturer of the measurement you are providing.

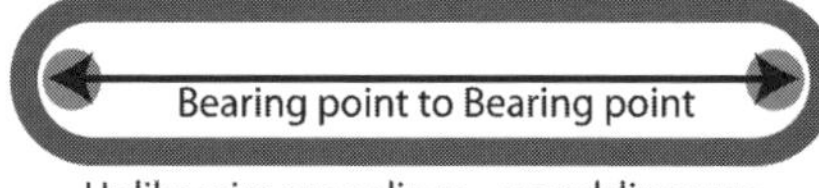
Bearing point to Bearing point

Unlike wire rope slings - roundslings are measured from bearing point to bearing point.

About capacity tables ...

Capacity tables are only intended to provide typical sling capacities for pre-job planning activities only. Always select and use all slings by the rated capacity (WLL) shown on the sling ID tag as provided by the manufacturer.

To determine the rating on wire-rope slings, the size, type and grade of rope must be known. It's best to always take a conservative approach with unknown wire rope slings and use the lowest ratings.

Current OSHA regulations (Slings- 1910.184) do not require identification tags on wire-rope slings (ASME B30.9 does require ID tags, but ASME is not a federal law). However OSHA does state that the safe working load must never be exceeded.

EIPS - MS

6x19 and 6x36 class
1-part wire rope sling

FIBER CORE

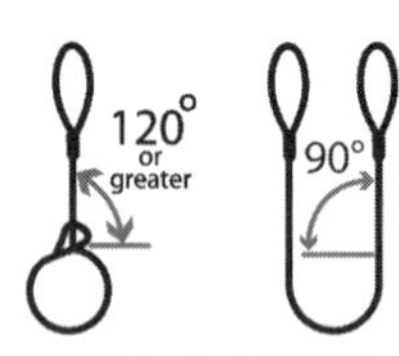
120°
or
greater
90°

SIZE	VERTICAL	CHOKER	BASKET
1/4"	0.56	0.42	1.1
5/16"	0.87	0.66	1.7
3/8"	1.2	0.94	2.5
7/16"	1.7	1.3	3.4
1/2"	2.2	1.6	4.4
9/16"	2.7	2.1	5.5
5/8"	3.4	2.6	6.8
3/4"	4.8	3.7	9.7
7/8"	6.6	5	13
1"	8.3	6.4	17
1-1/8"	10	8.1	21
1-1/4"	13	9.9	26

Values listed in U.S. tons

ASME B30.9

Rated capacities are based on:
1) a design factor of 5:1
2) basket hitches with a D/d ratio of 25
3) pin diameter no larger than the natural eye width
4) the sling diameter smaller than the pin diameter

IMPORTANT NOTICE: See pages 19 - 21 about the use of these capacity tables !

EIPS - MS

6x19 and 6x36 class
1-part wire rope sling

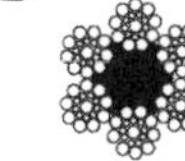

FIBER CORE

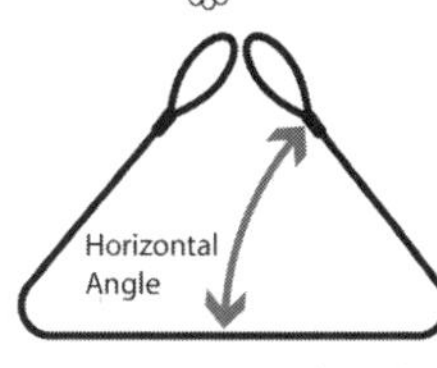
Horizontal
Angle

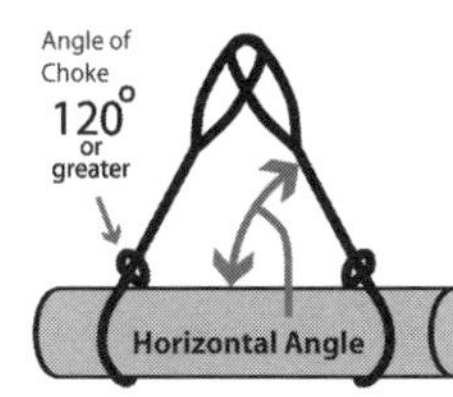
Angle of
Choke
120°
or
greater
Horizontal Angle

Basket			2-Chokers	
60°	30°	*SIZE*	60°	30°
0.97	0.56	**1/4"**	0.73	0.42
1.5	0.87	**5/16"**	1.1	0.66
2.2	1.2	**3/8"**	1.6	0.94
2.9	1.7	**7/16"**	2.2	1.3
3.8	2.2	**1/2"**	2.9	1.6
4.8	2.7	**9/16"**	3.6	2.1
5.9	3.4	**5/8"**	4.5	2.6
8.4	4.8	**3/4"**	6.3	3.7
11	6.6	**7/8"**	8.6	5
14	8.3	**1"**	11	6.4
18	10	**1-1/8"**	14	8.1
22	13	**1-1/4"**	17	9.9

Values listed in U.S. tons

ASME B30.9

Rated capacities are based on:
1) a design factor of 5:1
2) basket hitches with a D/d ratio of 25
3) pin diameter no larger than the natural eye width
4) the sling diameter smaller than the pin diameter
5) angle of choke is greater than 120 degrees

IMPORTANT NOTICE: See pages 19 - 21 about the use of these capacity tables !

EIPS - HT

6x19 and 6x36 class
1-part wire rope sling

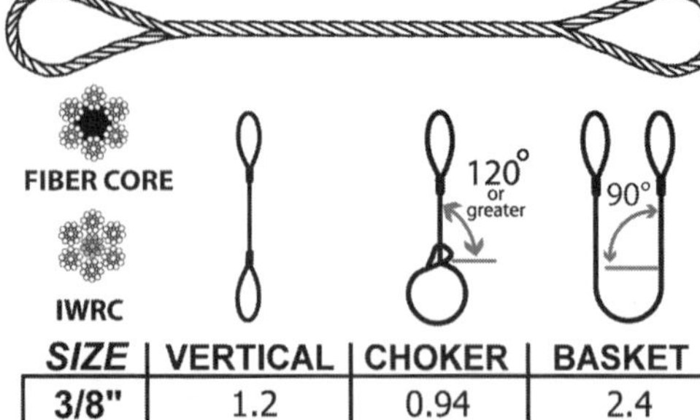
FIBER CORE
IWRC
120° or greater
90°

SIZE	VERTICAL	CHOKER	BASKET
3/8"	1.2	0.94	2.4
7/16"	1.6	1.3	3.2
1/2"	2.0	1.6	4.0
9/16"	2.5	2.1	5.0
5/8"	3.1	2.6	6.2
3/4"	4.3	3.7	8.6
7/8"	5.7	5	11
1"	7.4	6.4	15
1-1/8"	9.3	8.1	19
1-1/4"	11	9.9	23
1-3/8"	14	12	27
1-1/2"	16	14	32
1-5/8"	19	16	38
1-3/4"	22	19	44
1-7/8"	25	22	50
2"	28	25	56
2-1/8"	32	28	63
2-1/4"	35	31	70

Values listed in U.S. tons

ASME B30.9

Rated capacities are based on:
1) a design factor of 5:1
2) basket hitches with a D/d ratio of 25
3) pin diameter no larger than the natural eye width
4) the sling diameter smaller than the pin diameter

IMPORTANT NOTICE: See pages 19 - 21 about the use of these capacity tables !

EIPS - HT

6x19 and 6x36 class
1-part wire rope sling

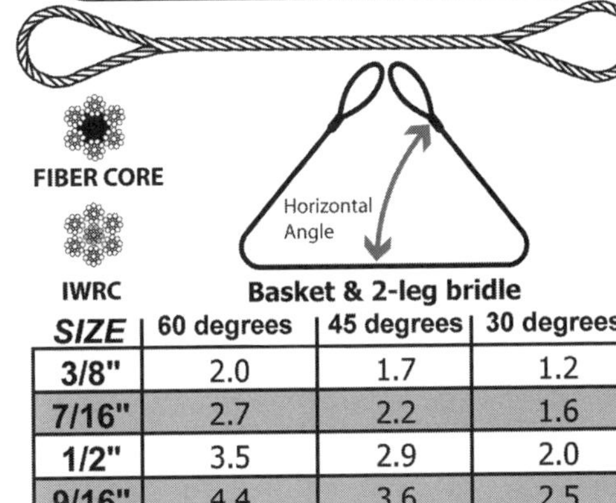
FIBER CORE
IWRC
Horizontal
Angle

Basket & 2-leg bridle

SIZE	60 degrees	45 degrees	30 degrees
3/8"	2.0	1.7	1.2
7/16"	2.7	2.2	1.6
1/2"	3.5	2.9	2.0
9/16"	4.4	3.6	2.5
5/8"	5.3	4.4	3.1
3/4"	7.4	6.1	4.3
7/8"	9.8	8	5.7
1"	13	10	7.4
1-1/8"	16	13	9.3
1-1/4"	20	16	11
1-3/8"	24	19	14
1-1/2"	28	23	16
1-5/8"	33	27	19
1-3/4"	38	31	22
1-7/8"	43	35	25
2"	49	40	28
2-1/8"	55	45	32
2-1/4"	61	50	35

Values listed in U.S. tons

ASME B30.9

Rated capacities are based on:
1) a design factor of 5:1
2) basket hitches with a D/d ratio of 25
3) pin diameter no larger than the natural eye width
4) the sling diameter smaller than the pin diameter

IMPORTANT NOTICE: See pages 19 - 21 about the use of these capacity tables !

EIPS - MS

6x19 and 6x36 class
1-part wire rope sling

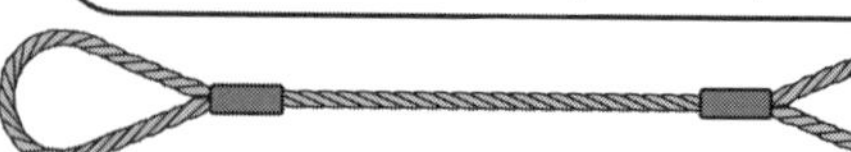

IWRC

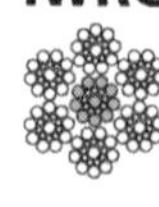

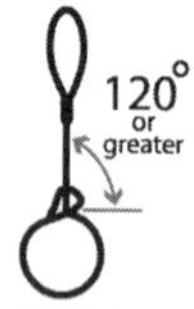
120°
or
greater

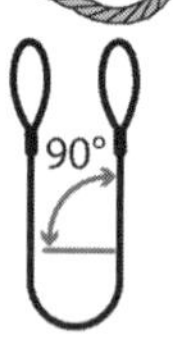
90°

SIZE	VERTICAL	CHOKER	BASKET
1/4"	0.65	0.48	1.3
5/16"	1	0.74	2
3/8"	1.4	1.1	2.9
7/16"	1.9	1.4	3.9
1/2"	2.5	1.9	5.1
9/16"	3.2	2.4	6.4
5/8"	3.9	2.9	7.8
3/4"	5.6	4.1	11
7/8"	7.6	5.6	15
1"	9.8	7.2	20
1-1/8"	12	9.1	24
1-1/4"	15	11	30
1-3/8"	18	13	36
1-1/2"	21	16	42
1-5/8"	24	18	49
1-3/4"	28	21	57
1-7/8"	32	24	64
2"	37	28	73

Values listed in U.S. tons

ASME B30.9

Rated capacities are based on:
1) a design factor of 5:1
2) basket hitches with a D/d ratio of 25
3) pin diameter no larger than the natural eye width
4) the sling diameter smaller than the pin diameter

IMPORTANT NOTICE: See pages 19 - 21 about the use of these capacity tables !

EIPS - MS

6x19 and 6x36 class
1-part wire rope sling

IWRC

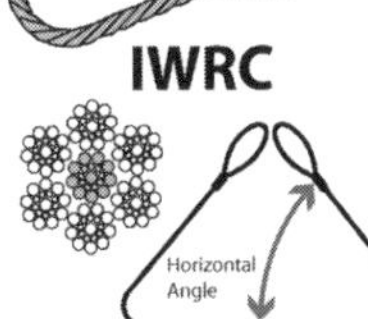
Horizontal
Angle

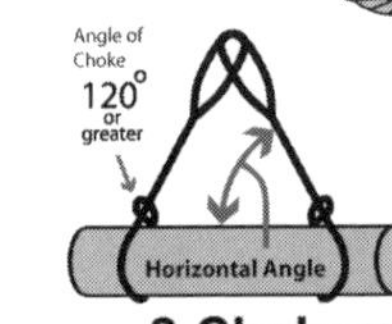
Angle of
Choke
120°
or
greater
Horizontal Angle

Basket 60°	Basket 30°	SIZE	2-Chokers 60°	2-Chokers 30°
1.1	0.65	**1/4"**	0.82	0.48
1.7	1	**5/16"**	1.3	0.74
2.5	1.4	**3/8"**	1.8	1.1
3.4	1.9	**7/16"**	2.5	1.4
4.4	2.5	**1/2"**	3.2	1.9
5.5	3.2	**9/16"**	4.1	2.4
6.8	3.9	**5/8"**	5	2.9
9.7	5.6	**3/4"**	7.1	4.1
13	7.6	**7/8"**	9.7	5.6
17	9.8	**1"**	13	7.2
21	12	**1-1/8"**	16	9.1
26	15	**1-1/4"**	19	11
31	18	**1-3/8"**	23	13
37	21	**1-1/2"**	28	16
42	24	**1-5/8"**	32	18
49	28	**1-3/4"**	37	21
56	32	**1-7/8"**	42	24
63	37	**2"**	48	28

Values listed in U.S. tons

ASME B30.9

Rated capacities are based on:
1) a design factor of 5:1
2) basket hitches with a D/d ratio of 25
3) pin diameter no larger than the natural eye width
4) the sling diameter smaller than the pin diameter

IMPORTANT NOTICE: See pages 19 - 21 about the use of these capacity tables !

EEIPS - MS

6x19 and 6x36 class 1-part wire rope sling

IWRC

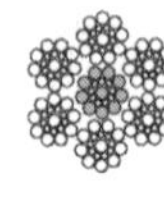

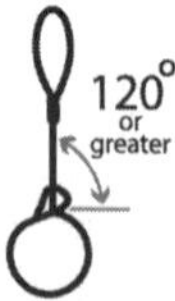
120°
or
greater

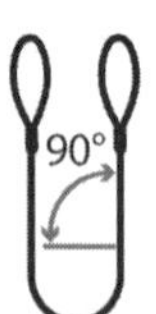
90°

SIZE	VERTICAL	CHOKER	BASKET
1/4"	0.71	0.52	1.4
5/16"	1.1	0.81	2.2
3/8"	1.6	1.2	3.2
7/16"	2.1	1.6	4.3
1/2"	2.8	2	5.5
9/16"	3.5	2.6	7
5/8"	4.3	3.2	8.6
3/4"	6.2	4.5	12
7/8"	8.3	6.1	17
1"	11	8	22

Values listed in U.S. tons

ASME B30.9

Rated capacities are based on:
1) a design factor of 5:1
2) basket hitches with a D/d ratio of 25
3) pin diameter no larger than the natural eye width
4) the sling diameter smaller than the pin diameter

IMPORTANT NOTICE: See pages 19 - 21 about the use of these capacity tables !

EEIPS - MS

6x19 and 6x36 class 1-part wire rope sling

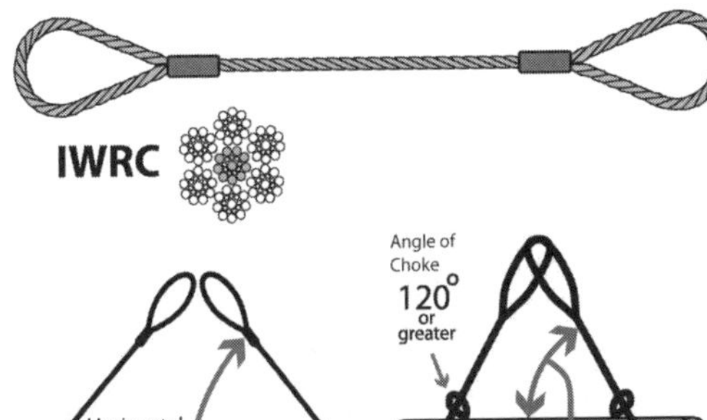
IWRC
Horizontal Angle
Angle of Choke
120° or greater
Horizontal Angle

Basket 60°	Basket 30°	SIZE	2-Chokers 60°	2-Chokers 30°
1.2	0.71	1/4"	0.9	0.52
1.9	1.1	5/16"	1.4	0.81
2.7	1.6	3/8"	2	1.2
3.7	2.1	7/16"	2.7	1.6
4.8	2.8	1/2"	3.5	2
6.1	3.5	9/16"	4.5	2.6
7.5	4.3	5/8"	5.5	3.2
11	6.2	3/4"	7.9	4.5
14	8.3	7/8"	11	6.1
19	11	1"	14	8

Values listed in U.S. tons

ASME B30.9

Rated capacities are based on:
1) a design factor of 5:1
2) basket hitches with a D/d ratio of 25
3) pin diameter no larger than the natural eye width
4) the sling diameter smaller than the pin diameter

IMPORTANT NOTICE: See pages 19 - 21 about the use of these capacity tables !

ROUNDSLINGS

Polyester Roundslings are available in 2 types:

EN (Endless)

EE (Eye-Eye)

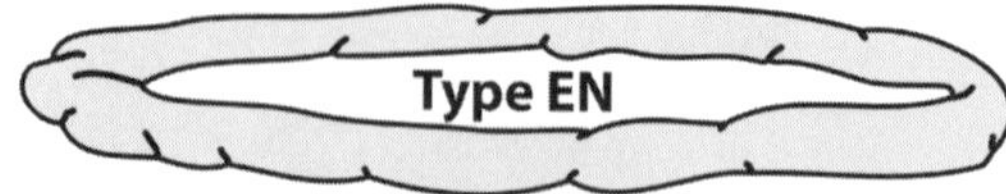
Type EN

Type EE

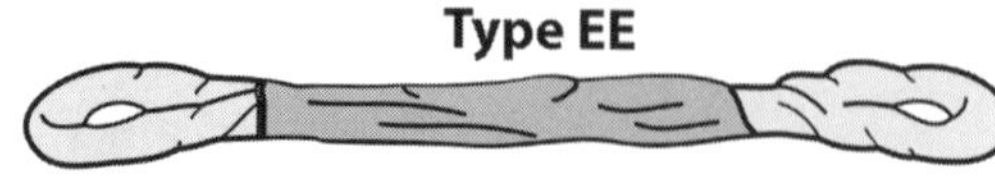

Elongation (stretch) for polyester rounslings is 3% at the full rated load

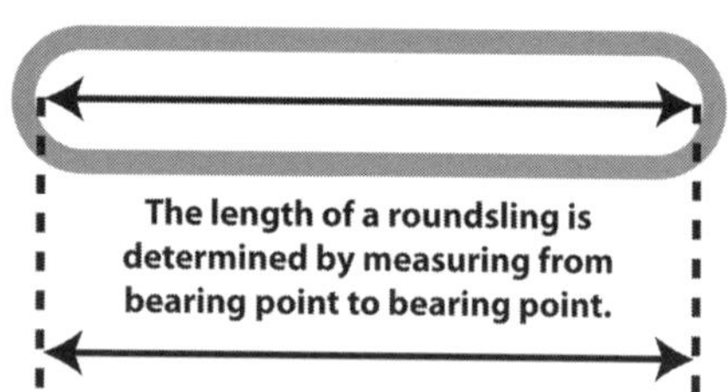
The length of a roundsling is determined by measuring from bearing point to bearing point.

Color guidelines for polyester roundsling covers are sometimes used to indicate the vertical rated capacity of roundslings; however, this is not required or always followed by manufacturers. Always use roundslings by the rated capacity as indicated on the tag - never rely on the color!

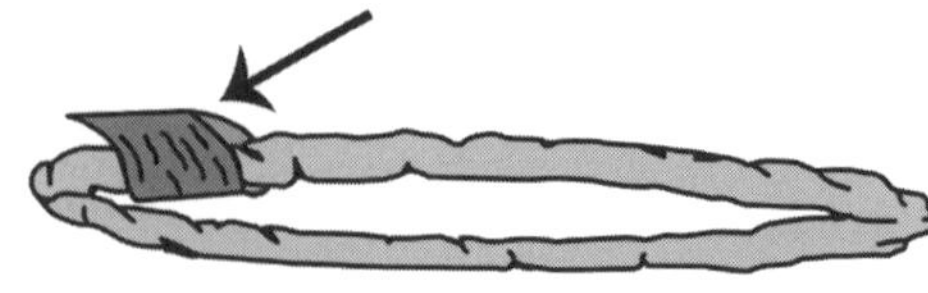

Misinterpretation of a basket hitch

When using a roundsling or any endless sling, make sure you choose the correct hitch AND the correct WLL for the sling.

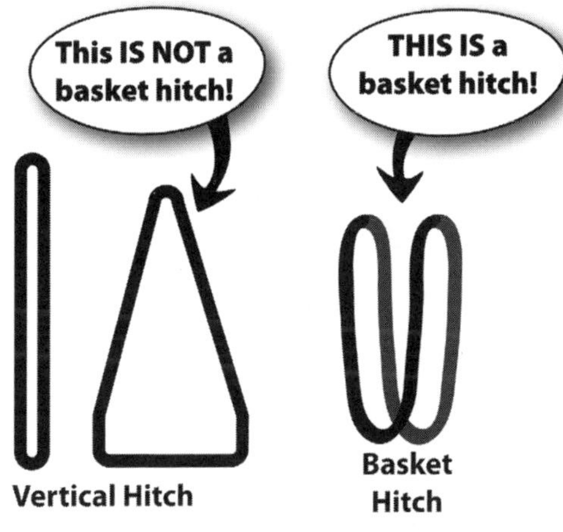
This IS NOT a basket hitch!
THIS IS a basket hitch!
Vertical Hitch
Basket Hitch

Just because the base of the hitch is wider than the top of the hitch DOES NOT imply that it's a basket hitch. A true basket hitch has twice the capacity of a vertical hitch because it has twice as many legs to support the load.

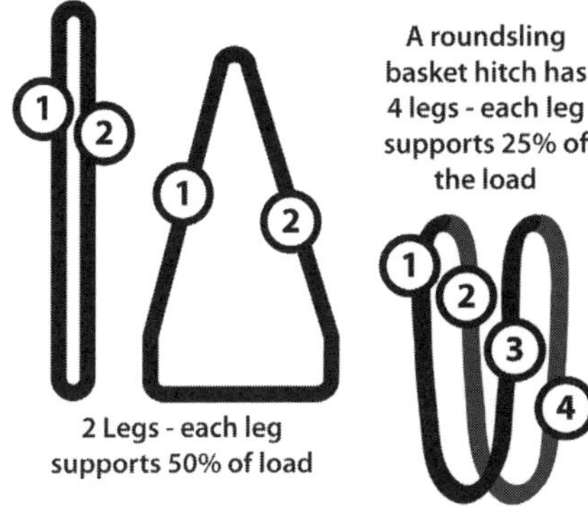
1
2
1
2
2 Legs - each leg supports 50% of load
A roundsling basket hitch has 4 legs - each leg supports 25% of the load
1
2
3
4

ROUNDSLINGS

Working Load Limits

For use with endless and eye-eye slings

WSTDA & ASME minimum values - Manufacturers may have different WLL's

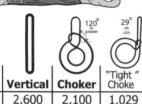

120° or greater

29° OR LESS

Size	Vertical	Choker	"Tight" Choke	Minimum Dia. Dec.	Minimum Dia. In.
1	2,600	2,100	1,029	0.5	1/2"
2	5,300	4,200	2,058	0.62	5/8"
3	8,400	6,700	3,283	0.75	3/4"
4	10,600	8,500	4,165	0.88	7/8"
5	13,200	10,600	5,194	1	1"
6	16,800	13,400	6,566	1.12	1- 1/8"
7	21,200	17,000	8,330	1.25	1- 1/4"
8	25,000	20,000	9,800	1.25	1- 1/4"
9	31,000	24,800	12,152	1.5	1- 1/2"
10	40,000	32,000	15,680	1.62	1- 5/8"
11	53,000	42,400	20,776	1.88	1- 7/8"
12	66,000	52,800	25,872	2.12	2- 1/8"
13	90,000	72,000	35,280	2.5	2- 1/2"

Maximum rated loads (WLL) are achieved by using the stock (pin) diameters as shown in the table. Stock diameters listed are from WSTDA-RS-2 (2005).

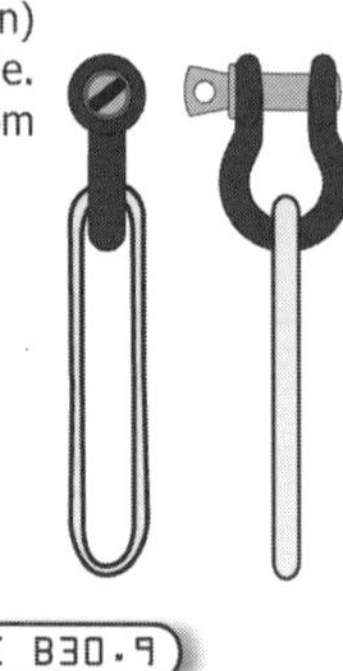

Never choke hardware!

ASME B30.9

The values shown are listed in US pounds and are based on ANSI B30.9 and WSTDA-RS1. Always consult the sling tag for the exact Working Load Limit of the sling being used.

ROUNDSLINGS

Working Load Limits

For use with endless and eye-eye slings

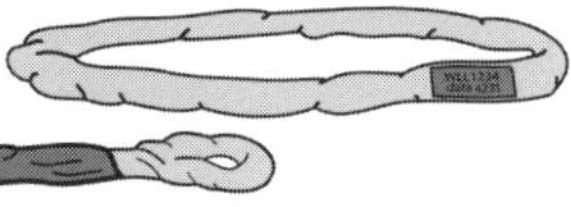

Size	2-Leg or Basket 90°	60°	45°	30°	Min Dia. In.
1	5,200	4,500	3,700	2,600	5/8"
2	10,600	9,200	7,500	5,300	7/8"
3	16,800	14,500	11,900	8,400	1"
4	21,200	18,400	15,000	10,600	1- 1/4"
5	26,400	22,900	18,700	13,200	1- 3/8"
6	33,600	29,100	23,800	16,800	1- 5/8"
7	42,400	36,700	30,000	21,200	1- 3/4"
8	50,000	43,300	35,400	25,000	1- 7/8"
9	62,000	53,700	43,800	31,000	2"
10	80,000	69,300	56,600	40,000	2- 3/8"
11	106,000	91,800	74,900	53,000	2- 3/4"
12	132,000	114,300	93,300	66,000	3"
13	180,000	155,900	127,300	90,000	3- 1/2"

Maximum rated loads (WLL) are achieved by using the stock (pin) diameters as shown in the table. Stock diameters listed are from WSTDA-RS-2 (2005).

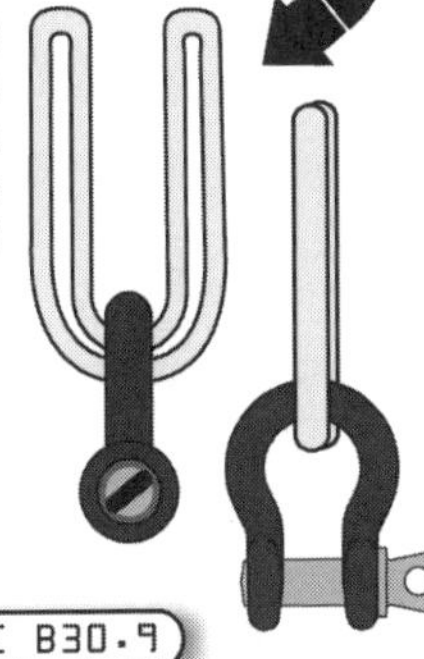

ASME B30.9

NYLON WEB SLINGS

Web Sling Types

TYPE 1 (TC)

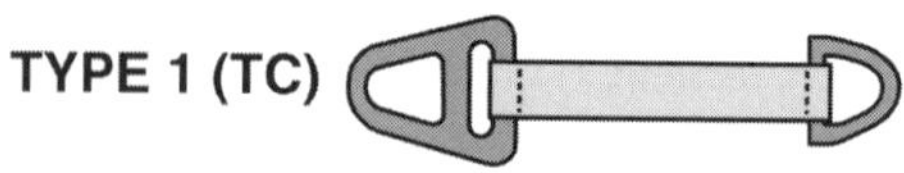

Slings have a triangle and choker fitting on either end. This is most commonly used in a choker hitch, but can also be used in basket and vertical hitches.

TYPE 2 (TT)

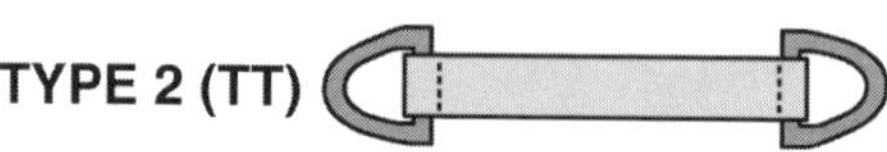

Slings have a triangle fitting on each end. They are normally used in a basket hitch. They CAN be used in a vertical hitch but they CANNOT be used in a choker hitch.

TYPE 3 (EE)

Flat Eye slings are very popular slings that can be used in all three types of hitches. They are easy to remove from beneath the load after the load is in place.

TYPE 4 (EE)

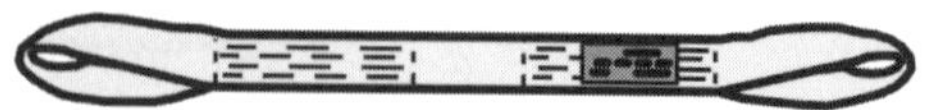

Twisted Eye slings are similar to Type 3 except the eyes are turned 90° to form a better choker hitch. This type of eye also nests together better when used in a basket hitch.

TYPE 5 (EN)

Endless slings can be used in all three types of hitches and wear points can be moved to increase sling life.

TYPE 6 (RE)

Reversed Eye slings have protective webbing over the body. This extra webbing reinforces the sling and protects it from wear.

There is no standard method of identifying web slings, however most nylon web slings will contain markings similar to those shown below.

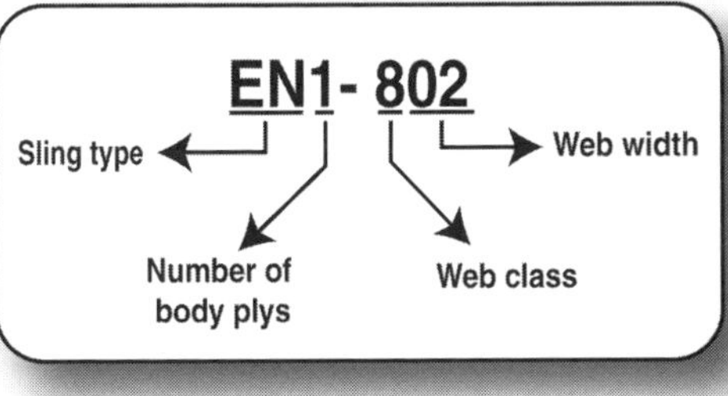
EN1- 802
Sling type
Web width
Number of body plys
Web class

1-ply WEB SLINGS EE

Working Load Limits

LIGHT DUTY

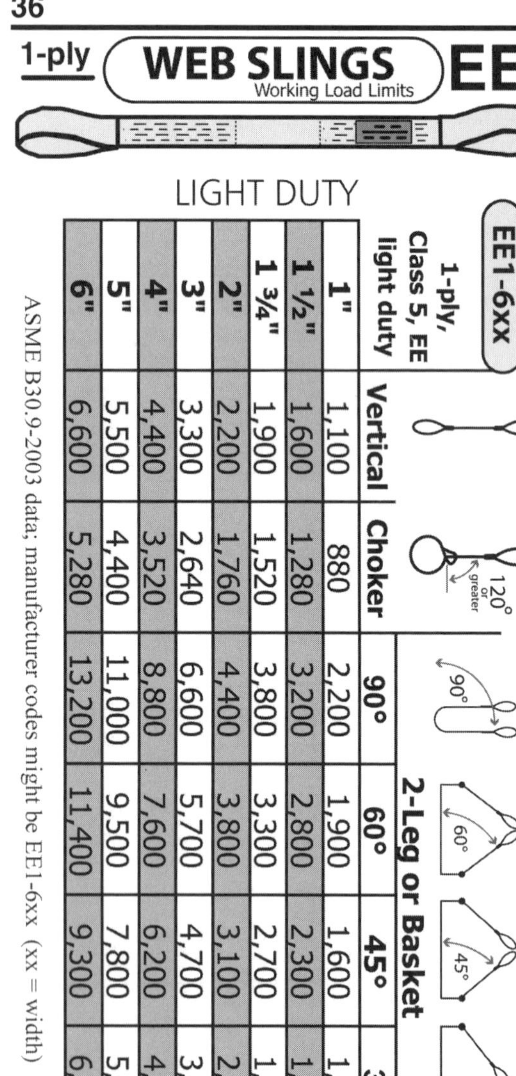

EE1-6xx 1-ply, Class 5, EE light duty	Vertical	Choker (120° or greater)	2-Leg or Basket 90°	2-Leg or Basket 60°	2-Leg or Basket 45°	2-Leg or Basket 30°
1"	1,100	880	2,200	1,900	1,600	1,100
1 1/2"	1,600	1,280	3,200	2,800	2,300	1,600
1 3/4"	1,900	1,520	3,800	3,300	2,700	1,900
2"	2,200	1,760	4,400	3,800	3,100	2,200
3"	3,300	2,640	6,600	5,700	4,700	3,300
4"	4,400	3,520	8,800	7,600	6,200	4,400
5"	5,500	4,400	11,000	9,500	7,800	5,500
6"	6,600	5,280	13,200	11,400	9,300	6,600

ASME B30.9-2003 data; manufacturer codes might be EE1-6xx (xx = width)

IMPORTANT NOTICE: See pages 19 - 21 about the use of these capacity tables !

SLINGS

2-ply WEB SLINGS EE

Working Load Limits

LIGHT DUTY

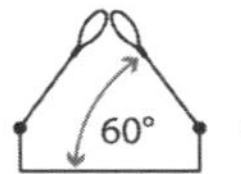
60°

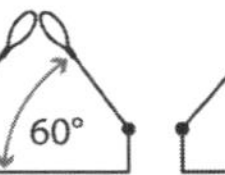
45°

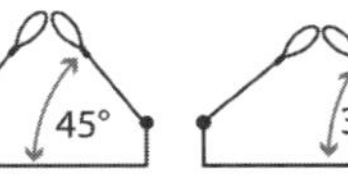
30°

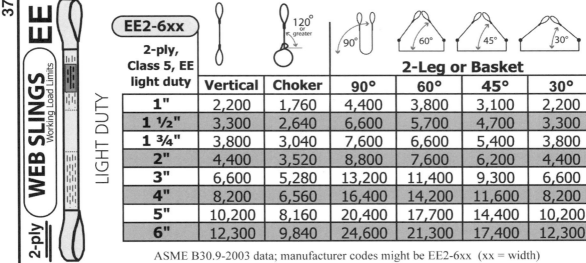

EE2-6xx 2-ply, Class 5, EE light duty	Vertical	Choker (120° or greater)	2-Leg or Basket 90°	60°	45°	30°
1"	2,200	1,760	4,400	3,800	3,100	2,200
1 ½"	3,300	2,640	6,600	5,700	4,700	3,300
1 ¾"	3,800	3,040	7,600	6,600	5,400	3,800
2"	4,400	3,520	8,800	7,600	6,200	4,400
3"	6,600	5,280	13,200	11,400	9,300	6,600
4"	8,200	6,560	16,400	14,200	11,600	8,200
5"	10,200	8,160	20,400	17,700	14,400	10,200
6"	12,300	9,840	24,600	21,300	17,400	12,300

ASME B30.9-2003 data; manufacturer codes might be EE2-6xx (xx = width)

1-ply WEB SLINGS EE

Working Load Limits

EE2-9xx

1-ply, Class 7, EE Heavy duty	Vertical	Choker (120° or greater)	2-Leg or Basket 90°	2-Leg or Basket 60°	2-Leg or Basket 45°	2-Leg or Basket 30°
1"	1,600	1,280	3,200	2,800	2,300	1,600
1 ½"	2,300	1,840	4,600	4,000	3,300	2,300
1 ¾"	2,700	2,160	5,400	4,700	3,800	2,700
2"	3,100	2,480	,6200	5,400	4,400	3,100
3"	4,700	3,760	9,400	8,100	6,600	4,700
4"	6,200	4,960	12,400	10,700	8,800	6,200
5"	7,800	6,240	15,600	13,500	11,000	7,800
6"	9,300	7,440	18,600	16,100	13,200	9,300
8"	11,750	9,400	21,150	18,300	15,000	11,750
10"	14,700	11,760	26,450	22,900	18,700	14,700
12"	17,650	14,120	31,750	27,500	22,400	17,650

11/08

IMPORTANT NOTICE: See pages 19 - 21 about the use of these capacity tables !

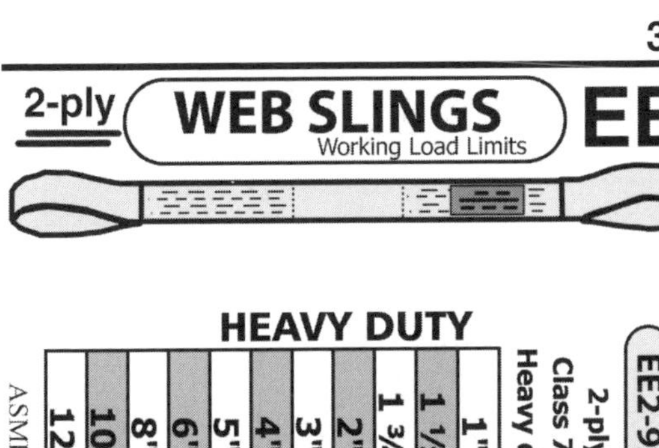
2-ply
WEB SLINGS
Working Load Limits
EE

HEAVY DUTY

EE2-9xx 2-ply, Class 7 EE Heavy duty	Vertical	Choker (120° or greater)	2-Leg or Basket 90°	60°	45°	30°
1"	3,100	2,480	6,200	5,400	4,400	3,100
1 ½"	4,700	3,760	9,400	8,100	6,600	4,700
1 ¾"	5,400	4,320	10,800	9,400	7,600	5,400
2"	6,200	4,960	12,400	10,700	8,800	6,200
3"	8,800	7,040	17,600	15,200	12,400	8,800
4"	11,000	8,800	22,000	19,100	15,600	11,000
5"	13,700	10,960	27,400	23,700	19,400	13,700
6"	16,500	13,200	33,000	28,600	23,000	16,500
8"	22,750	18,200	42,350	36,700	29,900	22,750
10"	28,400	22,720	52,900	45,800	37,400	28,400
12"	34,100	27,280	63,500	55,000	44,900	34,100

ASME B30.9-2003 data; manufacturer codes might be EE2-8xx or EE2-9xx (xx = width)

1-ply WEB SLINGS EN

Working Load Limits

LIGHT DUTY

EN1-60x

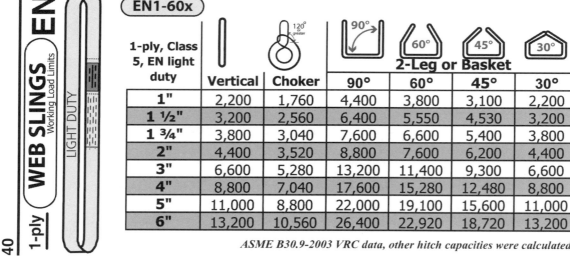

1-ply, Class 5, EN light duty	Vertical	Choker (120° or greater)	2-Leg or Basket 90°	60°	45°	30°
1"	2,200	1,760	4,400	3,800	3,100	2,200
1 1/2"	3,200	2,560	6,400	5,550	4,530	3,200
1 3/4"	3,800	3,040	7,600	6,600	5,400	3,800
2"	4,400	3,520	8,800	7,600	6,200	4,400
3"	6,600	5,280	13,200	11,400	9,300	6,600
4"	8,800	7,040	17,600	15,280	12,480	8,800
5"	11,000	8,800	22,000	19,100	15,600	11,000
6"	13,200	10,560	26,400	22,920	18,720	13,200

ASME B30.9-2003 VRC data, other hitch capacities were calculated

IMPORTANT NOTICE: See pages 19 - 21 about the use of these capacity tables !

2-ply WEB SLINGS EN

Working Load Limits

LIGHT DUTY

EN2-60x

90°

60°

45°

30°

2-ply, Class 5, EN light duty	Vertical	Choker (120° or greater)	2-Leg or Basket 90°	2-Leg or Basket 60°	2-Leg or Basket 45°	2-Leg or Basket 30°
1"	4,400	3,520	8,800	7,600	6,200	4,400
1 ½"	6,600	5,280	13,200	11,400	9,300	6,600
1 ¾"	7,600	6,080	15,200	13,000	10,000	7,600
2"	8,800	7,040	17,600	15,280	12,480	8,800
3"	13,200	10,560	26,400	22,920	18,720	13,200
4"	16,400	13,120	32,800	28,400	23,200	16,400
5"	20,400	16,320	40,800	35,400	28,900	20,400
6"	24,600	19,680	49,200	42,700	34,880	24,600

ASME B30.9-2003 VRC data, other hitch capacities were calculated

SLINGS | HARDWARE | FORMULAS | EXAMPLES | REFERENCE | STANDARDS | EXTRAS

1-ply WEB SLINGS EN

Working Load Limits

HEAVY DUTY

EN1-9XX

1-ply, Class 7, EN Heavy duty	Vertical	Choker (120° or greater)	2-Leg or Basket 90°	2-Leg or Basket 60°	2-Leg or Basket 45°	2-Leg or Basket 30°
1"	3,200	2,560	6,400	5,550	4,530	3,200
1 ½"	4,600	3,680	9,200	7,987	6,524	4,600
1 ¾"	5,400	4,320	10,800	9,376	7,658	5,400
2"	6,200	4,960	12,400	19,100	15,600	11,000
3"	9,400	7,520	18,800	16,322	13,331	9,400
4"	12,400	9,920	24,800	21,531	17,585	12,400
5"	15,600	12,480	31,200	27,087	22,124	15,600
6"	18,600	14,880	37,200	32,296	26,378	18,600
8"	21,150	16,920	42,300	36,724	29,995	21,150
10"	26,450	21,160	52,900	45,927	37,511	26,450
12"	31,750	25,400	63,500	55,130	45,027	31,750

ASME B30.9-2003 VRC data, other hitch capacities were calculated

IMPORTANT NOTICE: See pages 19 - 21 about the use of these capacity tables !

2-ply WEB SLINGS EN

Working Load Limits

HEAVY DUTY

EN2-9xx

2-ply, Class 7 EN Heavy duty	Vertical	Choker (120° or greater)	2-Leg or Basket 90°	60°	45°	30°
1"	6,200	4,960	12,400	10,765	8,793	6,200
1 ½"	9,400	7,520	18,800	16,322	13,331	9,400
1 ¾"	10,800	8,640	21,600	18,753	15,316	10,800
2"	12,400	9,920	24,800	21,531	17,585	12,400
3"	17,600	14,080	35,200	30,560	24,960	17,600
4"	22,000	17,600	44,000	38,200	31,200	22,000
5"	27,400	21,920	54,800	47,576	38,858	27,400
6"	33,000	26,400	66,000	57,300	46,800	33,000
8"	42,350	33,880	84,700	73,535	60,060	42,350
10"	52,900	42,320	105,800	91,854	75,022	52,900
12"	63,500	50,800	127,000	110,259	90,055	63,500

ASME B30.9-2003 VRC data, other hitch capacities were calculated

CHAIN SLINGS
Working Load Limits

Grade 80 Alloy

Grade 80 Alloy Steel Chain Slings SIZE	Vertical	Choker	2-Leg or Basket 60°	45°	30°
7/32" (5.5 mm)	2,100	1,700	3,600	3,000	2,100
9/32" (7 mm)	3,500	2,800	6,100	4,900	3,500
5/16" (8 mm)	4,500	3,600	7,800	6,400	4,500
3/8" (10 mm)	7,100	5,700	12,300	10,000	7,100
1/2" (13 mm)	12,000	9,600	20,800	17,000	12,000
5/8" (16 mm)	18,100	14,500	31,300	25,600	18,100
3/4" (20 mm)	28,300	22,600	49,000	40,000	28,300
7/8" (22 mm)	34,200	27,400	59,200	48,400	34,200
1" (26 mm)	47,700	38,200	82,600	67,400	47,700
1-1/4" (32 mm)	72,300	57,800	125,200	102,200	72,300

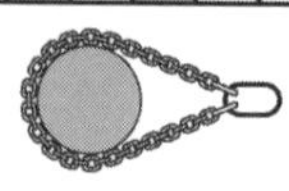

IMPORTANT NOTICE: See pages 19 - 21 about the use of these capacity tables !

CHAIN SLINGS
Working Load Limits

Grade 80 Alloy

3 or 4 Leg Bridle Sling Double Basket Sling

60°
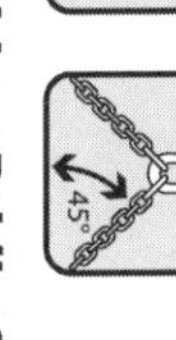
45°
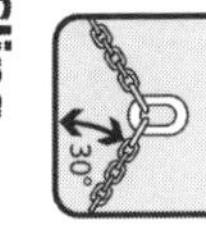
30°

Grade 80 Alloy Steel Chain Slings SIZE	60°	45°	30°
7/32" (5.5 mm)	5,500	4,400	3,200
9/32" (7 mm)	9,100	7,400	5,200
5/16" (8 mm)	11,700	9,500	6,800
3/8" (10 mm)	18,400	15,100	10,600
1/2" (13 mm)	31,200	25,500	18,000
5/8" (16 mm)	47,000	38,400	27,100
3/4" (20 mm)	73,500	60,000	42,400
7/8" (22 mm)	88,900	72,500	51,300
1" (26 mm)	123,900	101,200	71,500
1-1/4" (32 mm)	187,800	153,400	108,400

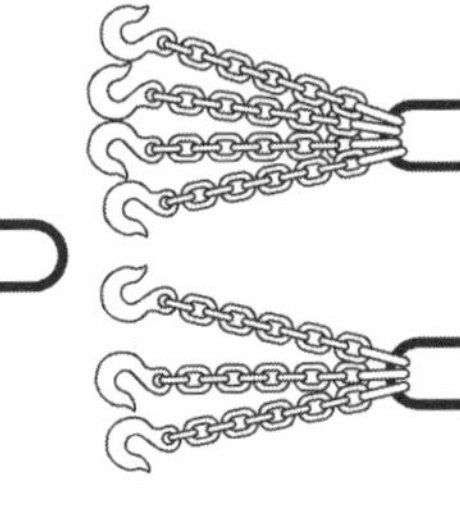

CHAIN SLINGS

Working Load Limits

Grade 100 Alloy

Grade 100 Alloy Steel Chain Slings			2-Leg or Basket		
SIZE	Vertical	Choker	60°	45°	30°
7/32" (5.5 mm)	2,700	2,100	4,700	3,800	2,700
9/32" (7 mm)	4,300	3,500	7,400	6,100	4,300
5/16" (8 mm)	5,700	4,500	9,900	8,100	5,700
3/8" (10 mm)	8,800	7,100	15,200	12,400	8,800
1/2" (13 mm)	15,000	12,000	26,000	21,200	15,000
5/8" (16 mm)	22,600	18,100	39,100	32,000	22,600
3/4" (20 mm)	35,300	28,300	61,100	49,900	35,300
7/8" (22 mm)	42,700	34,200	74,000	60,400	42,700

CHAIN SLINGS
Working Load Limits

Grade 100 Alloy

60°

45°

30°

Grade 100 Alloy Steel Chain Slings	3 or 4 Leg Bridle Sling Double Basket Sling		
SIZE	60°	45°	30°
7/32" (5.5 mm)	7,000	5,700	4,000
9/32" (7 mm)	11,200	9,100	6,400
5/16" (8 mm)	14,800	12,100	8,500
3/8" (10 mm)	22,900	18,700	13,200
1/2" (13 mm)	39,000	31,800	22,500
5/8" (16 mm)	58,700	47,900	33,900
3/4" (20 mm)	91,700	74,900	53,000
7/8" (22 mm)	110,900	90,600	64,000

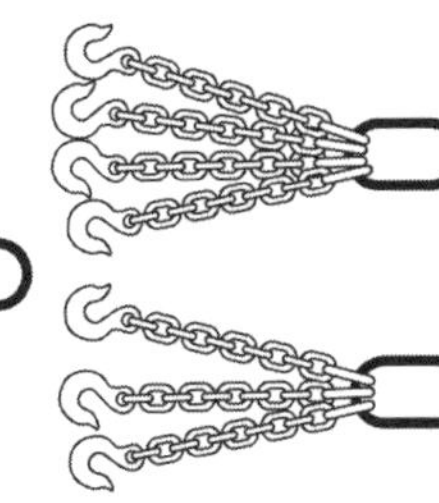

IMPORTANT NOTICE: See pages 19 - 21 about the use of these capacity tables !

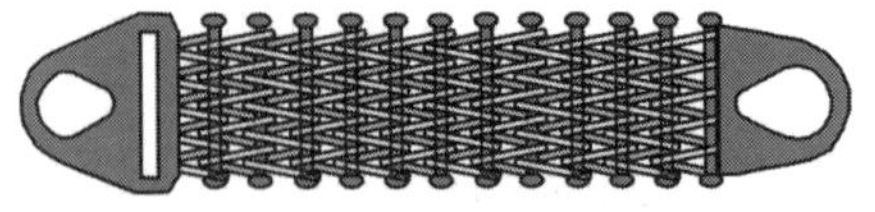

Metal Mesh, HEAVY DUTY 10 Ga., #35 Rated Capacities (in lbs)		
Sling Width (in inches)	Vertical or Choker	Vertical Basket
2	1,500	3,000
3	2,700	5,400
4	4,000	8,000
6	6,000	12,000
8	8,000	16,000
10	10,000	20,000
12	12,000	24,000
14	14,000	28,000
16	16,000	32,000
18	18,000	36,000
20	20,000	40,000

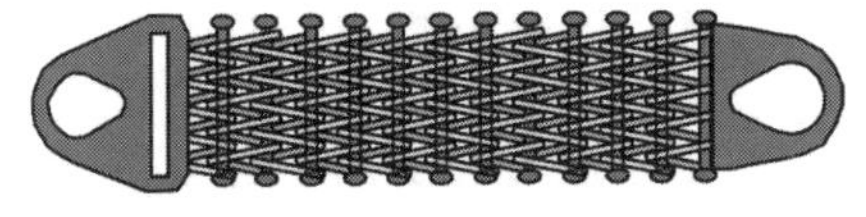

Metal Mesh, MEDIUM DUTY 12 Ga., #43 Rated Capacities (in lbs)		
Sling Width (in inches)	Vertical or Choker	Vertical Basket
2	1,350	2,700
3	2,000	4,000
4	2,700	5,400
6	4,500	9,000
8	6,000	12,000
10	7,500	15,000
12	9,000	18,000
14	10,500	21,000
16	12,000	24,000
18	13,500	27,000
20	1,500	30,000

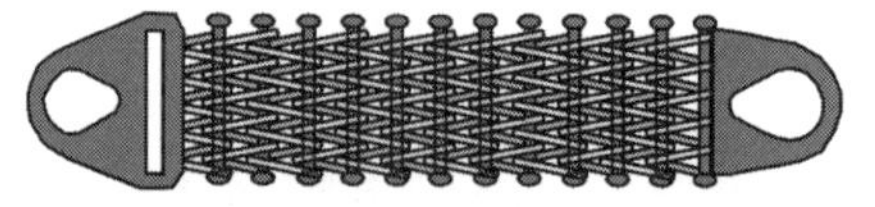

Metal Mesh, LIGHT DUTY 14 Ga., #59 Rated Capacities (in lbs)		
Sling Width (in inches)	Vertical or Choker	Vertical Basket
2	900	1,800
3	1,400	2,800
4	2,000	4,000
6	3,000	6,000
8	4,000	8,000
10	5,000	10,000
12	6,000	12,000
14	7,000	14,000
16	8,000	16,000
18	9,000	18,000
20	10,000	20,000

Twin-Path®

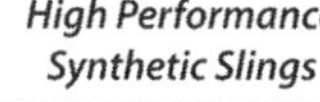
High Performance
Synthetic Slings

Classified as a roundsling by ASME, they have 2 separate "bundles" of core yarns that are wound in multiple parallel paths that support the load equally, thus earning the name "twin path".

Slingmax® Twin-Path® slings are the strongest and lightest on the market today.

Twin-Path® Extra slings contain a high-performance synthetic fiber known as K-Spec®, but they are sometimes nicknamed "Kevlar Slings". However the K-Spec fiber is not the same as Kevlar. They are two different materials, and slings manufactured from 100% Kevlar are no longer available.

Pound-for-pound these slings are five times stronger than steel at only 1/10th the weight of steel slings and with only 1% stretch at rated capacity.

Twin-Path® *High Performance Synthetic Slings*

TWIN-PATH®
SLING

Twin-Path® P/N [(1)]	WLL in lbs	
	Vertical	Choker
TPXC 1000	10,000	8,000
TPXC 1500	15,000	12,000
TPXC 2000	20,000	16,000
TPXC 2500	25,000	20,000
TPXC 3000	30,000	24,000
TPXC 4000	40,000	32,000
TPXC 5000	50,000	40,000
TPXC 6000	60,000	48,000
TPXC 7000	70,000	56,000
TPXC 8500	85,000	68,000
TPXC 10000	100,000	80,000
TPXC 12500	125,000	100,000
TPXC 15000	150,000	120,000
TPXC 17500	175,000	140,000
TPXC 20000	200,000	160,000
TPXC 25000	250,000	200,000
TPXC 27500	275,000	220,000
TPXC 30000	300,000	240,000
TPXC 40000	400,000	320,000
TPXC 50000	500,000	400,000

(1) Stock Number are for slings with a Covermax™ Cover

Data for the TWIN-PATH® EXTRA SLING WITH COVERMAX® and K-SPEC® Core Yarn TPXC has been provided by: SLINGMAX® Rigging Solutions (www.slingmax.com) Telephone: 610-485-8500

Twin-Path® *High Performance Synthetic Slings*

Basket Hitches - WLL in lbs

90° 60° 45°

Twin-Path® P/N (1)	90 deg	60 deg	45 deg
TPXC 1000	20,000	17,320	14,140
TPXC 1500	30,000	25,980	21,210
TPXC 2000	40,000	34,640	28,280
TPXC 2500	50,000	43,300	35,350
TPXC 3000	60,000	51,960	42,420
TPXC 4000	80,000	69,280	56,560
TPXC 5000	100,000	86,139	70,700
TPXC 6000	120,000	103,920	84,840
TPXC 7000	140,000	121,240	98,980
TPXC 8500	170,000	147,220	120,190
TPXC 10000	200,000	173,200	141,400
TPXC 12500	250,000	216,500	176,750
TPXC 15000	300,000	259,800	212,100
TPXC 17500	350,000	303,100	247,450
TPXC 20000	400,000	346,400	282,800
TPXC 25000	500,000	433,000	353,500
TPXC 27500	550,000	476,300	388,850
TPXC 30000	600,000	519,600	424,200
TPXC 40000	800,000	689,112	565,600
TPXC 50000	1,000,000	861,390	707,000

(1) Stock Number are for slings with a Covermax™ Cover

Data for the TWIN-PATH® EXTRA SLING WITH COVERMAX® and K-SPEC® Core Yarn TPXC has been provided by: SLINGMAX® Rigging Solutions (www.slingmax.com) Telephone: 610-485-8500

Twin-Path® *High Performance Synthetic Slings*

Component data table

VRC: Vertical Rated Capacity of the sling

Wide: The width of the sling

Pin: Minimum shackle stock/pin diameter to be used with this sling. Based on recommendation from Slingmax® Tech Bulletin #3 "Ratings based on straight pin diameter, one-half the sling width."

Wt: Approximate Weight in lbs. per ft (Measure sling from bearing point to bearing point)

Twin-Path® P/N	VRC	Wide	Pin	Wt
TPXC 1000	10,000	3"	1.5"	0.31
TPXC 1500	15,000	3"	1.5"	0.40
TPXC 2000	20,000	3"	1.5"	0.55
TPXC 2500	25,000	4"	2"	0.65
TPXC 3000	30,000	4"	2"	0.80
TPXC 4000	40,000	5"	2.5"	1.12
TPXC 5000	50,000	5"	2.5"	1.50
TPXC 6000	60,000	5"	2.5	1.60
TPXC 7000	70,000	6"	3"	1.66
TPXC 8500	85,000	6"	3"	1.85
TPXC 10000	100,000	6"	3"	2.20
TPXC 12500	125,000	8"	4"	3.00
TPXC 15000	150,000	8"	4"	3.36
TPXC 17500	175,000	10"	5"	4.00
TPXC 20000	200,000	10"	5"	4.37
TPXC 25000	250,000	11"	5.5"	5.50
TPXC 27500	275,000	11"	5.5"	6.90
TPXC 30000	300,000	13"	6.5"	7.50
TPXC 40000	400,000	14"	7"	8.60
TPXC 50000	500,000	16"	8"	11.0

Data for the TWIN-PATH® EXTRA SLING WITH COVERMAX® and K-SPEC® Core Yarn TPXC has been provided by: SLINGMAX® Rigging Solutions (www.slingmax.com) Telephone: 610-485-8500

Shackle Pin Protection

Graphics ©2007 Jerry Klinke

The pin area of a shackle can cause synthetic slings to be cut, and placing synthetic slings on the pin should be avoided.

There is usually a sharp edge where the threaded pin goes through the shackle ear.

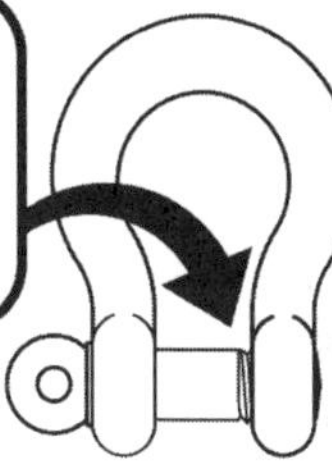

If the sling is exposed to this area under a load, the sling may be cut and fail.

If you must rig on the pin, you must protect shackle areas that are in contact with the sling.

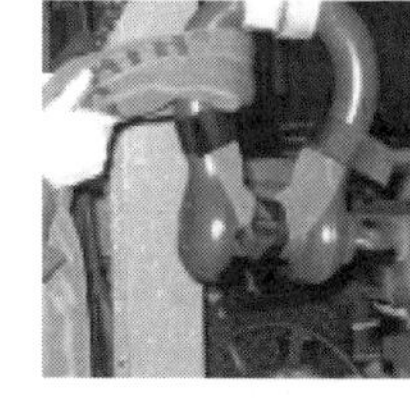

SLINGMAX® makes a Shackle Pin Pad (shown above) designed for this.

Shackle pin pads available at: SLINGMAX® Rigging Solutions (www.slingmax.com) Telephone: 610-485-8500

Twin-Path® Inspections

Twin-path slings are supplied with Tell-Tails and a Fiber Optic that assist in sling inspections.

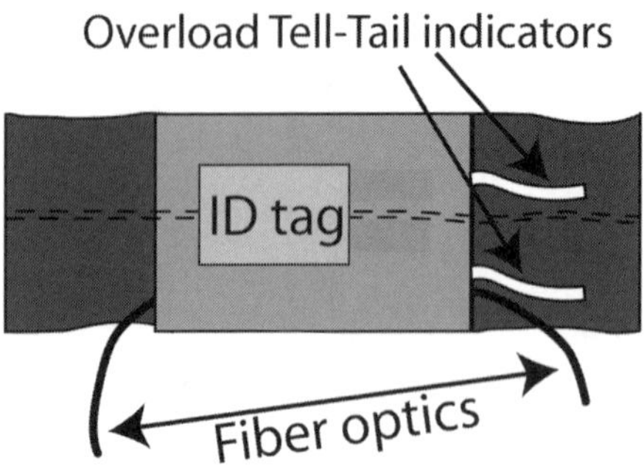
Overload Tell-Tail indicators
ID tag
Fiber optics

The **Tell-Tails are overload indicators** and will retract and eventually disappear if the sling is overloaded.

Tell-Tails should extend past the tag area of each sling. **If the Tell-Tails are not visible, remove the sling from service.** Send the sling to the manufacturer for repair evaluation.

The Fiber Optic determines if the interior core of the sling has suffered chemical, heat or crushing damage. If light does not pass from one end to the other, remove the sling from service and send to the manufacturer for repair evaluation.

In addition, these sling are considered a roundsling by ASME and the inspection criteria listed on page 60 should also be followed when inspecting these high performance slings.

Sling Inspections

OSHA

Safe operating practices (OSHA 1910.184) - Whenever any sling is used the following practices shall be observed:

1. Slings that are damaged or defective shall not be used.
2. Slings shall not be shortened with knots or bolts or other makeshift devices.
3. Sling legs shall not be kinked.
4. Slings shall not be loaded in excess of their rated capacities.
5. Slings used in a basket hitch shall have the loads balanced to prevent slippage.
6. Slings shall be securely attached to their loads.
7. Slings shall be padded or protected from the sharp edges of their loads.
8. Suspended loads shall be kept clear of all obstructions.
9. All employees shall be kept clear of loads about the be lifted and of suspended loads.
10. Hands or fingers shall not be placed between the sling and its load while the sling is being tightened around the load.
11 Shock loading is prohibited.
12. A sling shall not be pulled from under a load when the load is resting on the sling.

Inspections - Each day before being used, the sling and all fastenings and attachments shall be inspected for damage or defects by a competent person designated by the employer. Additional inspections shall be performed during sling use where service conditions warrant. Damaged or defective slings shall be immediately removed from service.

Wire Rope Slings (ASME B30.9) - A wire rope sling shall be removed from service if conditions such as the following are present:

1. Missing or illegible sling identification.
2. Broken Wires:
   - For strand-laid and single-part slings, ten randomly distributed broken wires in on rope lay, or five broken wires in one strand in one rope lay.
   - For cable-laid slings, 20 broken wires per lay.
   - For six-part braided slings, 20 broken wires per braid
   - For eight-part braided slings, 40 broken wires per braid.
3. Severe localized abrasion or scraping.
4. Kinking, crushing, bird caging, or any other damage resulting in damage to the rope structure.
5. Evidence of heat damage
6. End attachments that are cracked, deformed, or worn to the extent that the strength of the sling is substantially affected.
7. Severe corrosion of the rope, end attachments, or fittings.
8. for hooks, removal criteria at stated in ASME B30.10.
9. Other conditions, including visible damage, that cause doubt as to the continued use of the sling.

Nylon Web Slings (ASME B30.9) - A synthetic webbing sling shall be removed from service if conditions such as the following are present:

1. Missing or illegible sling identification.
2. Acid or caustic burns.
3. Melting or charring of any part of the sling.
4. Holes, tears, cuts, or snags.
5. Broken or worn stitching in load bearing splices.
6. Excessive abrasive wear.
7. Knots in any part of the sling.
8. Discoloration and brittle or stiff areas on any part of the sling, which may mean chemical or ultraviolet/sunlight damage.
9. Fitting that are pitted, corroded, cracked, bent, twisted, gouged, or broken.
10. For hooks, removal criteria as stated in ASME B30.10
11. Other conditions, including visible damage, that cause double as to the continued use of the sling.

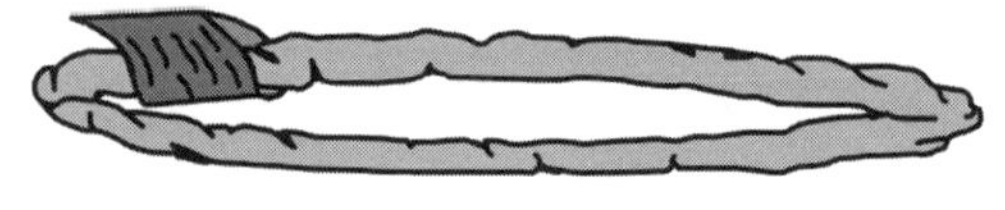

Polyester Round Slings (ASME B30.9) - A synthetic round sling shall be removed from service if conditions such as the following are present:

1. Missing or illegible sling identification.
2. Acid or caustic burns.
3. Evidence of heat damage.
4. Holes, tears, cuts, abrasive wear, or snags that expose the core yarns.
5. Broken or damaged core yarns.
6. Weld splatter that exposes core yarns.
7. Round slings that are knotted.
8. Discoloration and brittle or stiff areas on any part of the slings, which may mean chemical or ultraviolet/sunlight damage.
9. Fitting that are pitted, corroded, cracked, bent twisted, gouged, or broken.
10. For hooks, removal criteria as stated in ASME B30.10
11. Other conditions, including visible damage, that cause double as to the continued use of the sling.

Alloy Steel Chain Slings(ASME B30.9) - An alloy steel chain sling shall be removed from service if conditions such as the following are present:

1. Missing or illegible sling identification.
2. Cracks or breaks
3. Excessive wear, nicks, or gouges.
4. Stretched chain links or components
5. Bent, twisted, or deformed chain links or components.
6. Evidence of hear damage.
7. Excessive pitting or corrosion.
8. Lack of ability of chain or components to hinge (articulate) freely.
9. Weld splatter.
10. For hooks, removal criteria as stated in ASME B30.10
11. Other conditions, including visible damage, that cause double as to the continued use of the sling.

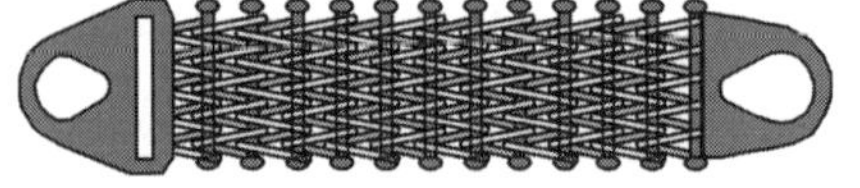

Wire Mesh Slings (ASME B30.9) - A metal mesh sling shall be removed from service if conditions such as the following are present:

1. Missing or illegible sling identification.
2. Broken weld or a broken brazed joint along the sling edge
3. Broken wire in any part of the mesh.
4. Reduction in wire diameter of 25% due to abrasion or 15% due to corrosion.
5. Lack of flexibility due to distortion of the mesh.
6. Distortion of the choker fitting so the depth of the slot is increased by more that 10%
7. Distortion of either end fitting so the width of the eye opening is decreased by more than 10%
8. A 15% reduction of the original cross-sectional area of any point around the hook opening of the end fitting.
9. Visible distortion of either end fitting out of its plane.
10. Cracked end fitting.
11. Slings in which the spirals are locked or without free articulation shall not be used.
12. Fitting that are pitted, corroded, cracked, bent, twisted, gouged, or broken.
13. Other conditions, including visible damage, that cause doubt as to the continued use of the sling.

For additional information, please refer to the OSHA and ASME standards.

Comparisons

Here are some typical prices for different types of slings. They have been grouped as close to the same rated capacity and length as possible. Prices were obtained in May 2007 and represent a typical cost, although many rigging dealers may have different prices on certain items.

SLING	Size	VRC	Price
Chain sling	9/32"	3500 lbs	$ 250.00
Wire Rope	3/8"	2800 lbs	$ 28.00
Web sling	1" wide	2400 lbs	$ 15.00
Roundsling	N/A	2600 lbs	$ 16.00
Mesh Sling	2" wide	2300 lbs	$ 280.00

Note: All slings are 4 feet in length

TwinPath® Slings

Size (weight)	VRC	Price
TPXC 1000 (8 lbs)	10,000 lbs	$ 594.00
TPXC 3000 (14 lbs)	30,000 lbs	$1,400.00

Wire Rope Slings

Size (weight)	VRC	Price
3/4" dia (26 lbs)	11,200 lbs	$ 112.00
1-1/4" dia (83 lbs)	30,000 lbs	$ 355.00

Note: All slings are 20 feet in length

VRC = Vertical Rated Capacity

	Design factor
Chain sling	**4:1**
Wire Rope	**5:1**
Web sling	**5:1**
Roundsling	**5:1**
Mesh Sling	**5:1**
TwinPath® Sling	**5:1**

Basic Terminology

Sling - an assembly which connects the load to the lifting equipment.

Eye - a fabricated loop, normally at the end of a sling, used as an attachment or choke point.

Leg - the extending portion of a sling used in a basket hitch or one extension of a sling with multiple parts.

Fitting - a general term for a piece of lifting hardware such as a hook, oblong, pear link, coupling, etc.

Reach - the working length of a lifting sling when pulled taut. It is measured from the load bearing point at one end of the sling, to the load bearing point at the opposite end.

Hitch - the way the sling is fastened to or around a load.

Sling Angle - the horizontal angle between the sling leg and the load, when pulled taut.

Working Load Limit - the maximum static load permitted by the manufacturer. (The terms "rated capacity" and "working load limit" are commonly used interchangeably.)

HARDWARE

The hardware used with rigging applications is just as important as the slings and wire ropes being used. Just as with slings, hardware that is used in angular rigging situations will experience additional loads.

All hardware is rated as if a straight, linear tension is applied. The Working Load Limits (W.W.L.) are reduced drastically when equipment is used improperly or at angles other than designed and rated for.

Shackles
Common Types

Anchor Shackles

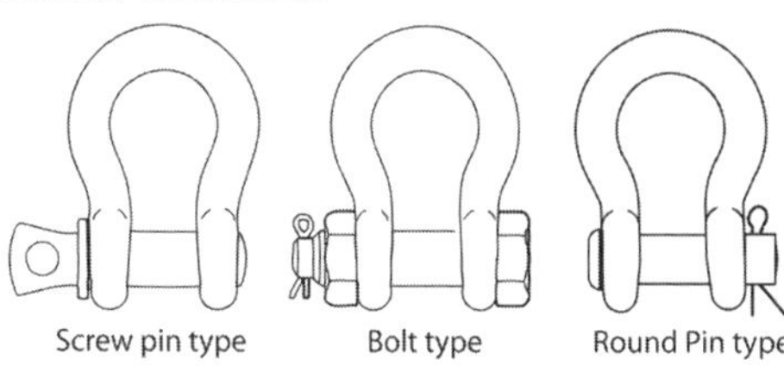

Screw pin type   Bolt type   Round Pin type

Chain Shackles

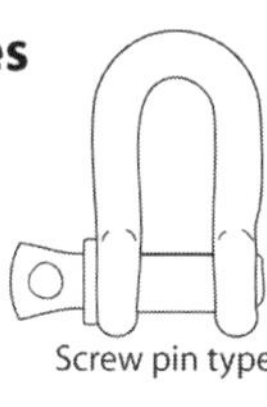

Screw pin type

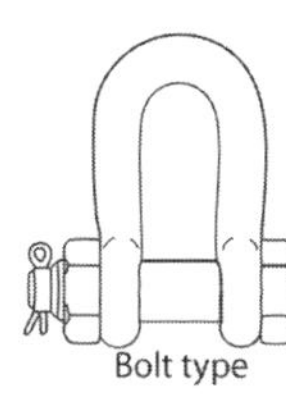

Bolt type

Synthetic Sling Shackles

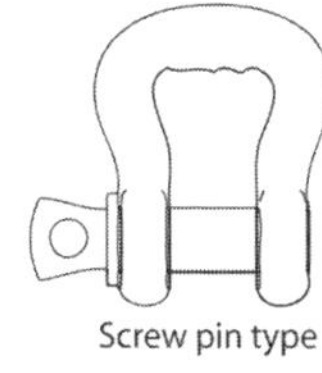

Screw pin type

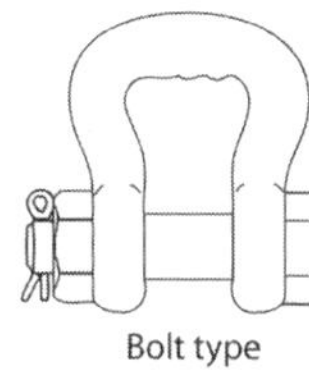

Bolt type

Wide Body Shackles

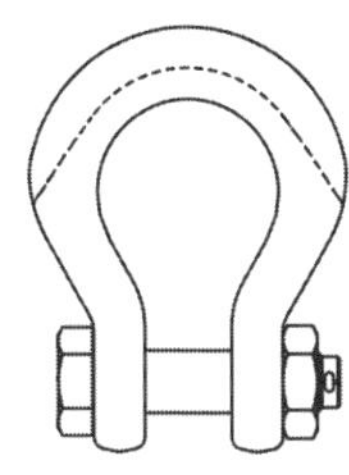

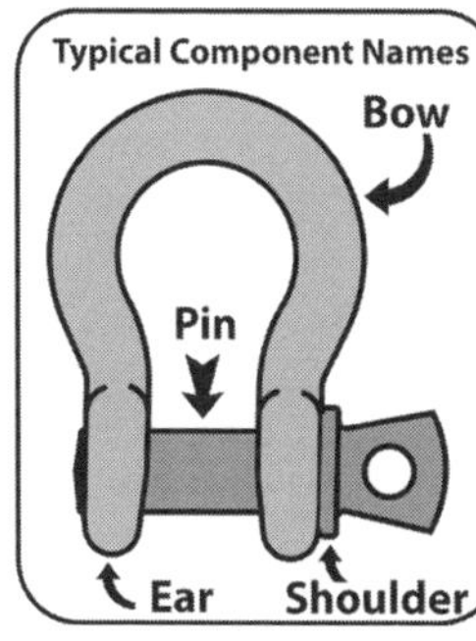
Typical Component Names
Bow
Pin
Ear
Shoulder

Swivel Hoist Rings
Common Types

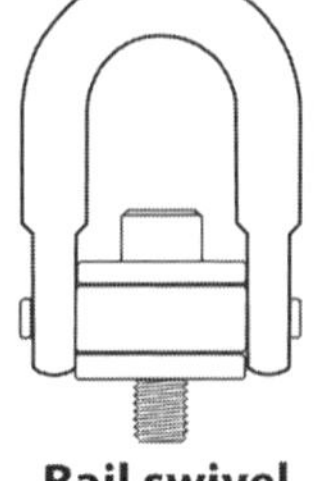

Bail swivel

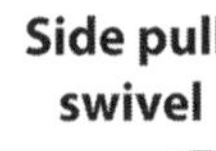

Side pull swivel

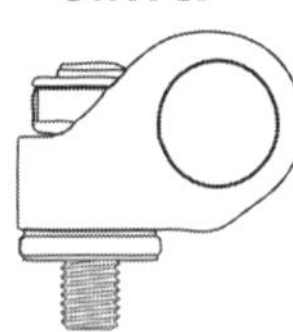

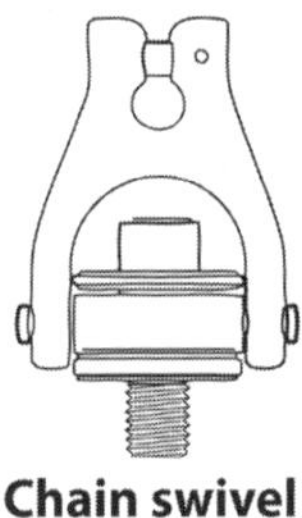

Chain swivel

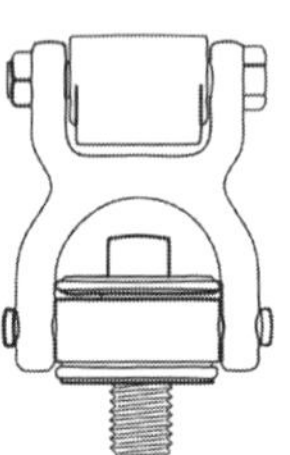

Webbing swivel

Typical Component Names

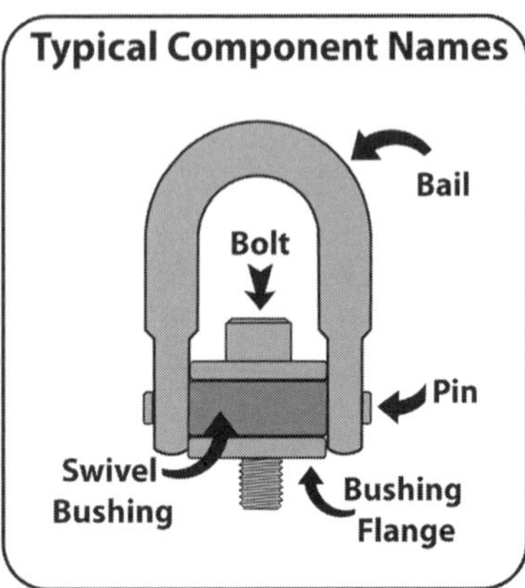
Bail
Bolt
Pin
Swivel Bushing
Bushing Flange

Hooks
Common Types

Some commonly used hooks for lifting/rigging operations

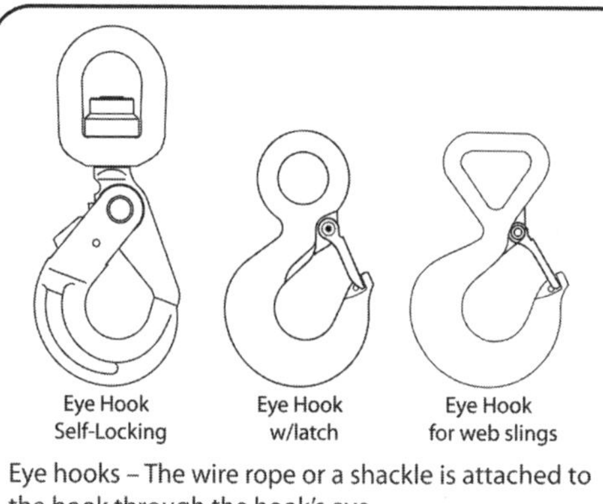

Eye Hook Self-Locking | Eye Hook w/latch | Eye Hook for web slings

Eye hooks – The wire rope or a shackle is attached to the hook through the hook's eye.

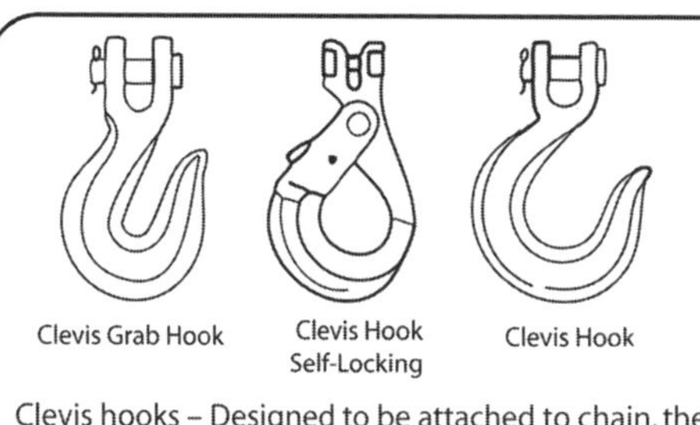

Clevis Grab Hook | Clevis Hook Self-Locking | Clevis Hook

Clevis hooks – Designed to be attached to chain, the link is secured by the pin in the clevis design.

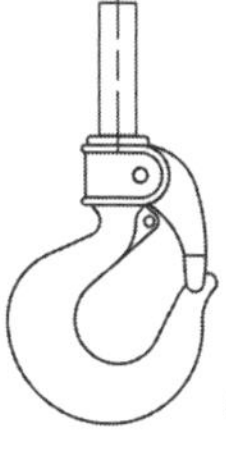

Shank hooks – Shank hooks are specifically design to swivel 360°. The one shown has a Self-Closing Gate Latch (sometimes called a Bullard hook).

Refer to ASME B30.10 "Hooks" for additional information.

Identification

Per ASME B30.26

NEW shackle bodys shall be marked to show:
(a) name or trademark of manufacturer
(b) rated load
(c) size

NEW shackle pins shall be marked to show:
(a) name or trademark of manufacturer
(b) grade, material type, or load rating

Swivel hoist ring shall be marked to show:
(a) name or trademark of manufacturer
(b) rated load
(c) torque value

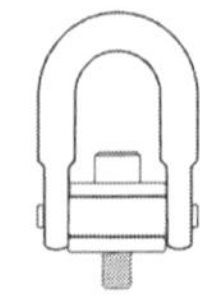

Turnbuckles, eyebolts, and eye nuts shall be marked to show:
(a) name or trademark of manufacturer
(b) size or rated load
(c) grade for alloy eyebolts

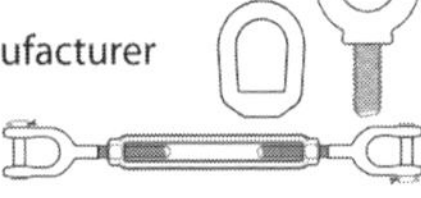

NEW wedge socket body and wedge shall be marked to show:
(a) name or trademark of manufacturer
(b) size
(c) model, if required to match wedge to body

NEW Wire rope clip saddles shall be marked to show:
(a) name or trademark of manufacturer
(b) size

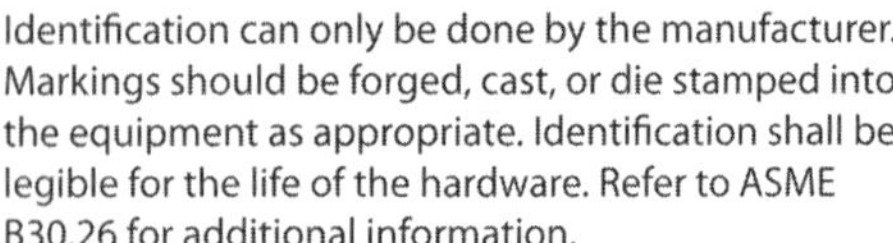

Identification can only be done by the manufacturer. Markings should be forged, cast, or die stamped into the equipment as appropriate. Identification shall be legible for the life of the hardware. Refer to ASME B30.26 for additional information.

ASME Consensus Standards are usually more rigorous than state and federal OSHA requirements, compliance is voluntary unless otherwise required.

Hardware Inspection

Per ASME B30.26

Initial Inspection*:
Prior to use, all new rigging hardware shall be inspected.

Frequent Inspection*:
A visual inspection shall be performed each day before the rigging hardware is used.

Periodic Inspection*:
A complete inspection of the rigging hardware shall be performed by a designated person at an interval not to exceed one year. The frequency is be based on; the frequency of use, service conditions, and other experience gained.

**Written records are not required.*

Removal Criteria
Rigging hardware shall be removed from service if damage such as the following is visible:

- missing or illegible identification or markings
- heat damage, weld spatter, arc strikes, excessive pitting, corrosion, nicks or gouges
- bent, twisted, distorted, stretched, elongated,cracked, or broken load-bearing components
- reduction (stretching) at any point around the body or pin, incomplete pin engagement, or excessive thread damage
- or any other conditions that cause doubt as to the continued use of the rigging hardware

Hardware temperature limits:

Swivel hoist rings: MAX 400°F (204°C) MIN -20°F (-29°C)
Carbon steel eyebolts: MAX 275°F (135°C) MIN 30°F (-1°C)
Shackles: MAX 400°F (204°C) MIN -40°F (-40°C)
Wire rope clips: MAX 400°F (204°C) MIN -40°F (-40°C)
Wedge sockets: MAX 400°F (204°C) MIN -4°F (-20°C)
Steel links & rings: MAX 400°F (204°C) MIN -40°F (-40°C)
Rigging blocks: MAX 150°F (66°C) MIN 0°F (-18°C)

Requirements shown are abridged and DO NOT address all the specific requirements for each type of rigging hardware. Refer to ASME B30.26 for a complete listing of these requirements.

The Capacity tables shown in this book are only intended to provide typical hardware capacities for pre-job planning activities.

Be aware than the hardware capacities are different depending on each sling manufacturer!

For example: the capacity for a 1" shackle can range from 8.5 tons to 16 tons depending on the type, grade and manufacturer.

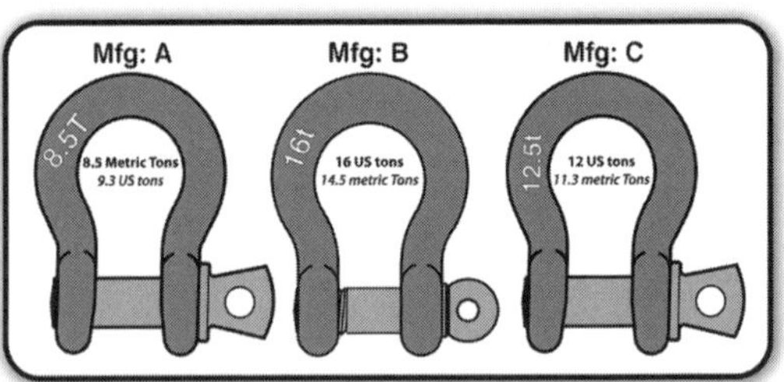
Mfg: A
8.5T
8.5 Metric Tons
9.3 US tons
Mfg: B
16t
16 US tons
14.5 metric Tons
Mfg: C
12.5t
12 US tons
11.3 metric Tons

Never assume that all rigging hardware is the same!

<u>Always select and use all rigging hardware by the rated capacity (WLL) as provided by the manufacturer.</u>

Wedge Socket Installation

The construction industry uses wedge sockets extensively because they attach easily to a wire rope. In applying the socket, the live rope should lead out of the socket in a straight line.

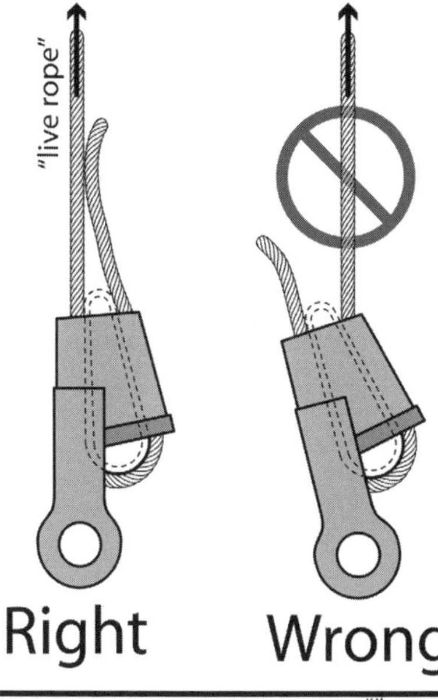
"live rope"
Right
Wrong

The tail length of the dead end of the wire rope should never be less than 6 inches. For standard wire rope the tail length should be a minimum of 6 wire rope diameters. For rotation resistant ropes the minimum length should be 20 rope diameters.

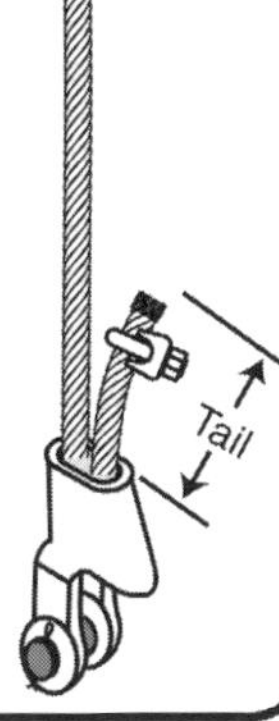
Tail

Wire rope clips used in conjunction with wedge sockets shall be attached only to the unloaded (dead end) of the rope - ASME B30.5-2004

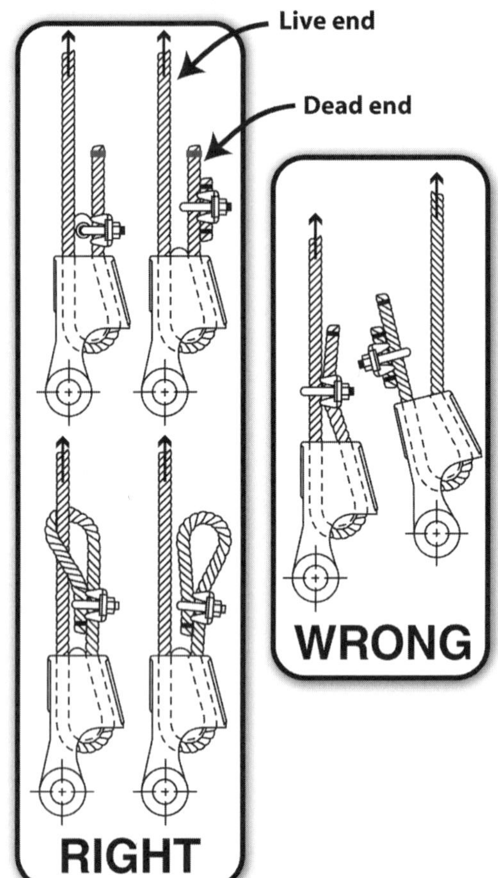
Live end
Dead end
RIGHT
WRONG

NOTE: Rotation Resistant Rope (not shown) requires additional precautions. Ensure that the dead end is brazed or seized before inserting into the wedge socket to prevent core slippage.

When using wedge sockets:

- Make sure that a sudden jolt or impact does not dislodge a wedge.
- When installing wire rope, always pre-load the wedge with wire rope in place.
- Check frequently to re-tighten or re-position as necessary.
- Make allowance for the crimping effect common with all types of wedge sockets. Experience shows that it will reduce the Working Load of a line by 20 percent.

Inserting line into the wedge socket

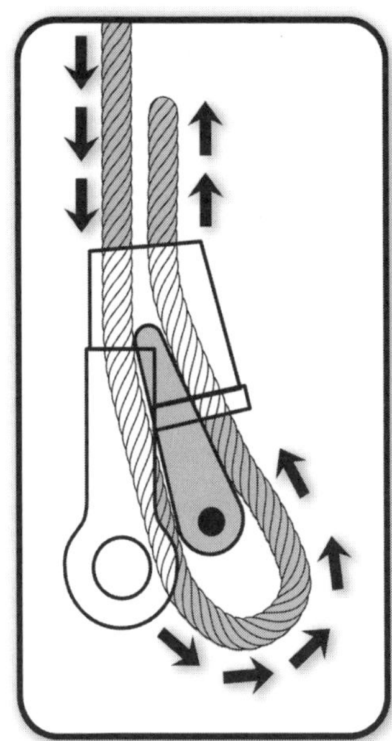

Make sure that the wedge is correct for the wire rope size! Generally the size is shown on the side wedge and the socket body.

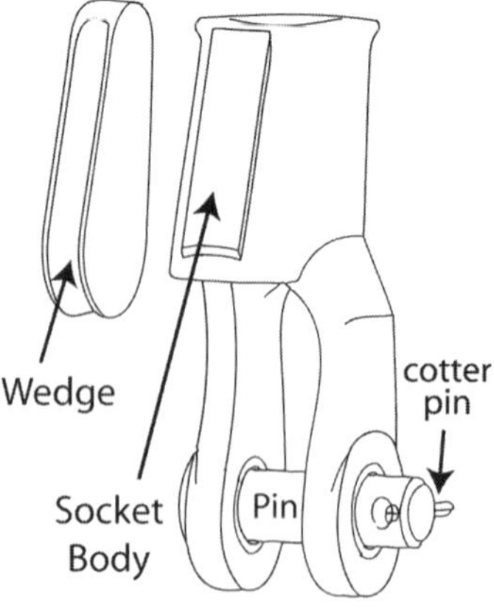
Wedge
Socket
Body
Pin
cotter
pin

Components

Warnings:

- Loads may slip or fall if the wedge socket is not properly installed.
- Read and understand the manufacturer's instructions before installing the wedge socket.
- Do not side load the wedge socket.
- Do not interchange one manufacturer's wedge socket, wedge, or pin with another manufacturer's product.
- Apply the first load to fully seat the wedge and wire rope in the socket. This load should be of equal or greater weight than loads expected during use/lifting.
- Do not interchange wedges between metric and U.S. devices or between sockets of different sizes.
- Before installing a wedge socket on plastic coated or plastic impregnated wire rope, consult the wedge socket manufacturer, wire rope manufacturer, or a qualified person.

Eyebolts

ANSI B18.15*

Forged Alloy Shoulder Machinery Eyebolts

NOTE: These values are only valid for hardware that was manufactured to meet ANSI/ASME B18.15 Standards.

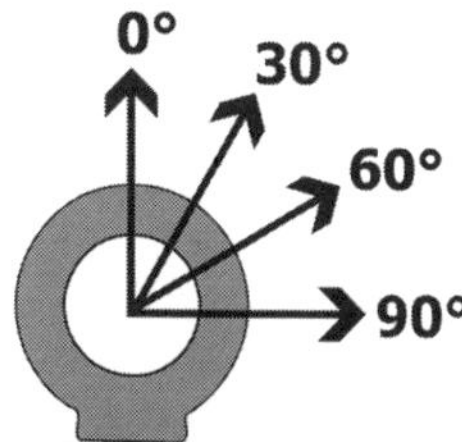
0°
30°
60°
90°

Eyebolts from an unknown manufacturer should NEVER be used for overhead lifting!

ALWAYS VERIFY WLL's with the equipment manufacturer before use.

SIZE	0°	30°	60°	90°
1/4"	400	75	NR	NR
5/16"	680	210	NR	NR
3/8"	1,000	400	220	180
7/16"	1,380	530	330	260
1/2"	1,840	850	520	440
9/16"	2,370	1,160	700	570
5/8"	2,940	1,410	890	740
3/4"	4,340	2,230	1,310	1,140
7/8"	6,000	2,960	1,910	1,630
1"	7,880	3,850	2,630	2,320
1-1/8"	9,920	4,790	3,840	3,390
1-1/4"	12,600	6,200	4,125	3,690
1-1/2"	18,260	9,010	6,040	5,460
1-3/4"	24,700	12,100	8,250	7,370
2"	32,500	15,970	10,910	9,740

WLL values shown in pounds

** ANSI/ASME B18.15-1985 (Reaffirmed 1995, 2003)*

Eyebolts

Crosby® S-279

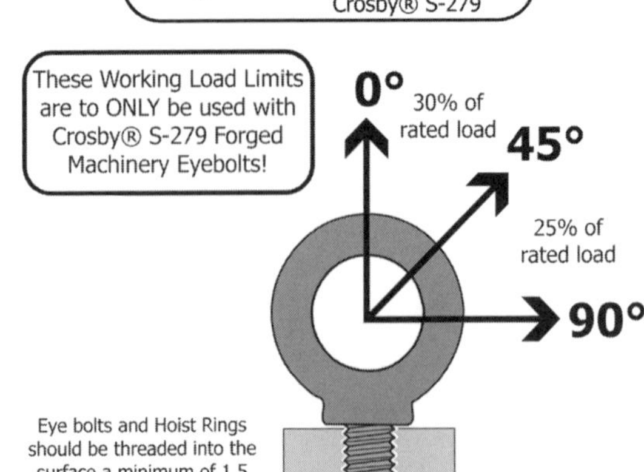
These Working Load Limits are to ONLY be used with Crosby® S-279 Forged Machinery Eyebolts!
0°
30% of rated load
45°
25% of rated load
90°

Eye bolts and Hoist Rings should be threaded into the surface a minimum of 1.5 times the thread diameter (ASME B30.26-2.9.4.2)

SIZE	0°	45°	90°
1/4" x 20	650	195	163
5/16" x 18	1,200	360	300
3/8" x 16	1,550	465	388
1/2" x 13	2,600	780	650
5/8" x 11	5,200	1,560	1,300
3/4" x 10	7,200	2,160	1,800
7/8" x 9	10,600	3,180	2,650
1" x 8	13,300	3,990	3,325
1-1/4" x 7	21,000	6,300	5,250
1-1/2" x 6	24,000	7,200	6,000

Angular lifts will significantly lower working load limits and should be avoided whenever possible. If an angular lift is required, a properly seated eye bolt must be used. Loads should always be applied to eye bolts in the plane of the eye, not at an angle to this plane. Contact the manufacturer for detailed information.

Eyebolts

Machinery Eye Bolts by Chicago Hardware

Straight pull ONLY

CHICAGO

5

Plain

SIZE	Dia x TPI	Straight Pull WLL
1	1/4" x 20	500
2	5/16" x 18	900
3	3/8" x 16	1,400
4	7/16" x 14	2,000
5	1/2" x 13	2,600
6	9/16" x 12	3,200
7	5/8" x 11	4,000
8	3/4" x 10	6,000
9	7/8" x 9	7,000
10	1" x 8	9,000
11	1-1/8" x 7	12,000
12	1-1/4" x 7	15,000
14	1-1/2" x 6	21,000

Working load limits for eye bolts are based on a straight vertical lift in a gradually increasing manner. Angular lifts (Shoulder Pattern only) will significantly lower working load limits and should be avoided whenever possible. Angular lifts (Shoulder Pattern only) must never be more than a 45° pull.

Reprinted from the Chicago Hardware 2003 catalog, available online at: www.chicagohardware.com

Straight pull

45° pull

CHICAGO

23

Shoulder

SIZE	Dia x TPI	Straight Pull WLL	45° Pull WLL
21	1/4" x 20	500	125
22	5/16" x 18	900	225
23	3/8" x 16	1,400	350
24	7/16" x 14	2,000	500
25	1/2" x 13	2,600	650
26	9/16" x 12	3,200	750
27	5/8" x 11	4,000	1,000
28	3/4" x 10	6,000	1,500
29	7/8" x 9	7,000	1,750
30	1" x 8	9,000	2,250
31	1-1/8" x 7	12,000	2,500
32	1-1/4" x 7	15,000	3,750
34	1-1/2" x 6	21,000	4,900

Eyebolts

Operating Practices

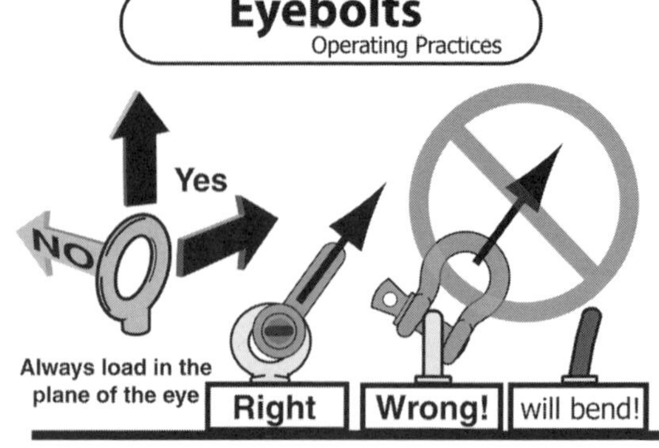
Yes
NO
Always load in the plane of the eye
Right
Wrong!
will bend!

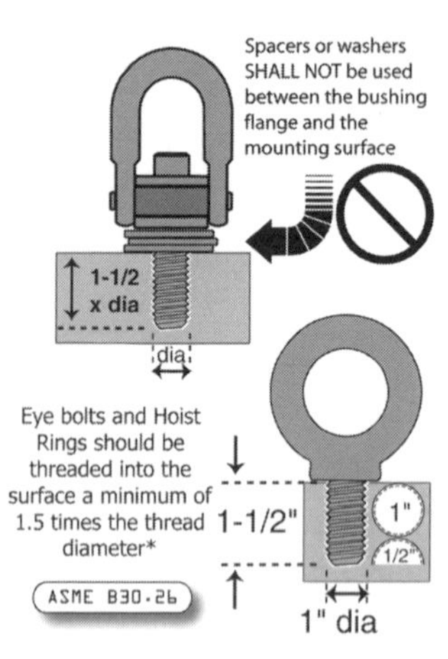
Spacers or washers SHALL NOT be used between the bushing flange and the mounting surface
1-1/2 x dia
dia
Eye bolts and Hoist Rings should be threaded into the surface a minimum of 1.5 times the thread diameter*
1-1/2"
1"
1/2"
1" dia
ASME B30.26

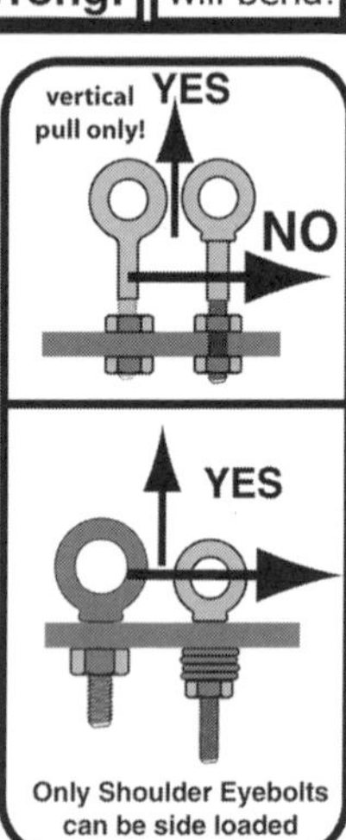
vertical pull only!
YES
NO
YES
Only Shoulder Eyebolts can be side loaded

* per ASME B30.26-2.9.4.2 "when used in a tapped blind hole, the effective thread length shall be at least 1-1/2 times the diameter of the bolt for engagement in steel ... For other thread engagements or in other materials, contact the eyebolt manufacturer or a qualified person."

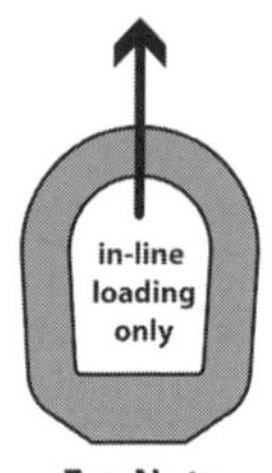
in-line loading only

Eye Nuts

Eyebolts

Alignment

Eye Bolt size (inches)	Shim Thickness Required to change rotation 90 degrees (inches)
1/4" x 20 tpi	0.0125
5/16" x 18 tpi	0.0139
3/8" x 16 tpi	0.0156
1/2" x 13 tpi	0.0192
5/8" x 11 tpi	0.0227
3/4" x 10 tpi	0.0250
7/8" x 9 tpi	0.0278
1" x 8 tpi	0.0312
1-1/4" x 7 tpi	0.0357
1-1/2" x 6 tpi	0.0417

tpi = threads per inch

When a shouldered eyebolt needs to be turned to align with the rigging, a shim is added to reposition the eyebolt and maintain tightness.

Swivel Hoist Rings

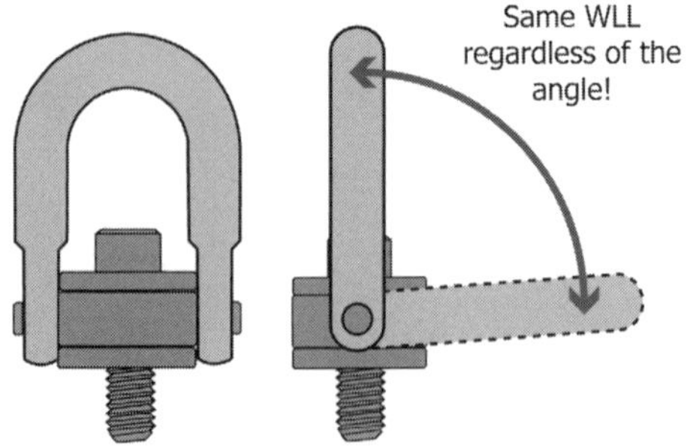
Same WLL
regardless of the
angle!

Key issues when using Hoist Rings:

- The threaded holes must be clean, and not damaged.
- The hoist ring must be fully seated with all the threads of the bolt fully engaged.
- Always torque hoist rings to the proper value.
- Tightening torque values shown are based upon threads being clean, dry and free of lubrication
- Bolt specifications are typically Grade 8 Alloy socket head cap screw (per ASTM A-574 standard) All values are based on threads that are UNC.
- When a Hoist Ring is installed with a nut, the nut must have full thread engagement and must meet the meet the manufacturers requirements.

Certain manufacturers may list different working load limits for this type of lifting equipment. Always consult the manufacturer for specific installation and use recommendations when using this type of equipment.

Swivel Hoist Rings

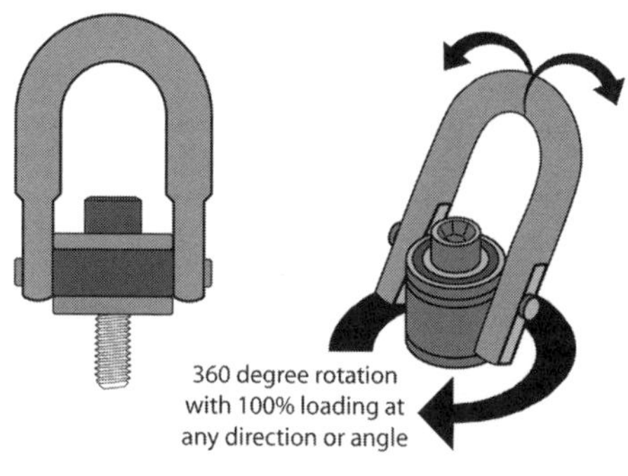
360 degree rotation
with 100% loading at
any direction or angle

Swivel Hoist Rings		
Size - thread	WLL (lbs)	Torque
5/16 x 18	800	7
3/8 x 16	1000	12
1/2 x 13	2500	28
5/8 x 11	4000	60
3/4 x 10	5000	100
7/8 x 9	8000	160
1 x 8	10000	230
1-1/4 x 7	15000	470
1-1/2 x 6	24000	800

Tightening torque values shown are based upon threads being clean, dry and free of lubrication

Swivel Hoist Rings

The Working Load Limit and torque value are typically stamped on the top of most hoist rings.

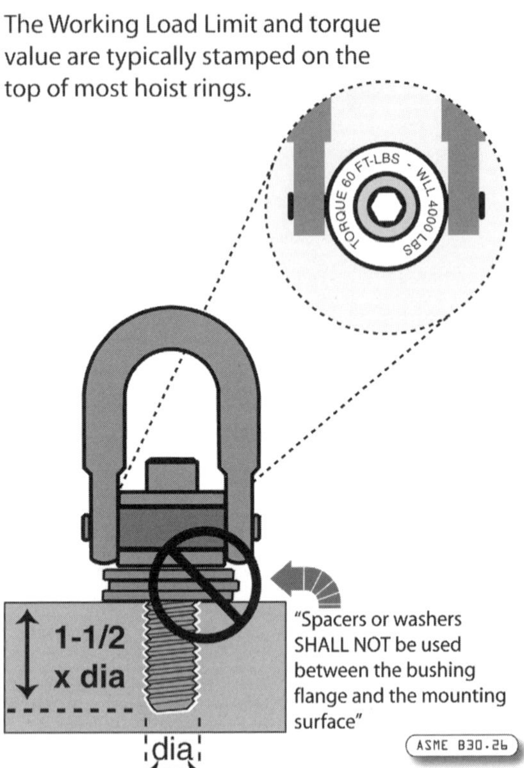
TORQUE 60 FT-LBS - WLL 4000 LBS
1-1/2 x dia
dia
"Spacers or washers SHALL NOT be used between the bushing flange and the mounting surface"
ASME B30.26

Eye bolts and Hoist Rings should be threaded into the surface a minimum of 1.5 times the thread diameter*

** per ASME B30.26-2.9.4.2 "when used in a tapped blind hole, the effective thread length shall be at least 1-1/2 times the diameter of the bolt for engagement in steel ... For other thread engagements or in other materials, contact the eyebolt manufacturer or a qualified person."*

Always consult the manufacturer for specific installation and use recommendations when using this type of equipment.

End Termination
strength efficiency

Wire rope attachments develop less than full strength of the wire rope. Always consult the manufacturer for the exact values of the equipment. These examples are based on typical equipment and installations.

Wedge Sockets
up to 80%

Spelter Sockets
up to100%

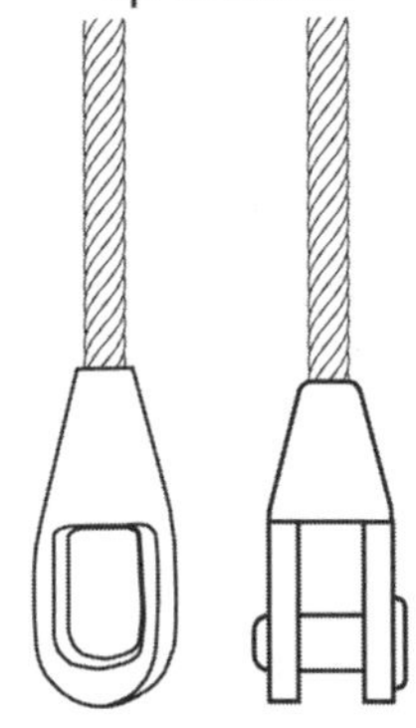

Wire Rope Clips

approximately 80% efficiency if properly installed

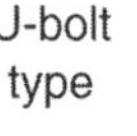

U-bolt type

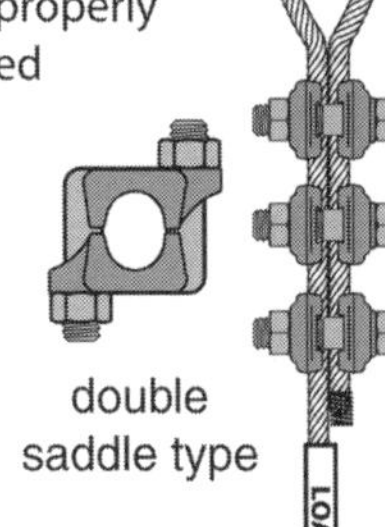

double saddle type

LOAD
LOAD

SLINGS HARDWARE FORMULAS EXAMPLES REFERENCE STANDARDS EXTRAS

Wire Rope Clips

NEVER saddle
a dead horse!

This common phrase is a way to help remember the correct way to install a wire rope clip.

It tells you that the U-bolt goes on the dead end of the rope – where crushing will not affect the strength of the rope.

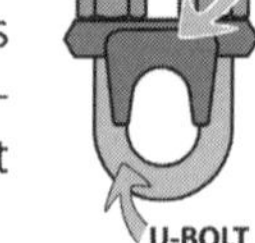
SADDLE
U-BOLT

Don't stagger wire rope clips!

LOAD

Even one incorrect clamp will reduce the capacity of the end termination.

Wire Rope
Connections

"Wire rope clips shall not be used to fabricate wire rope slings except where the application of slings prevents the use of prefabricated slings and where the specific application is designed by a qualified person."

ASME B30.9

What does this mean?
Basically, that you shouldn't fabricate wire rope slings for general repetitive daily use.
However - for special or one-time unusual lifts wire rope clips may be used to fabricate rigging including slings. This method should only be used in special cases, provided the wire rope clips' manufacturer's recommendations are followed explicitly.

Refer to AMSE B30 for additional information

"Eyes in wire rope bridles, slings, or bull wires shall not be formed by wire rope clips or knots."

OSHA 1926.251(c)(4)(iii)

Wire Rope Clips
Installation

STEP 1 Determine the amount of turn back required from the tables shown on the next page.

Apply first clip one base width from the dead end of the rope. Tighten nuts evenly and torque at the recommended value.

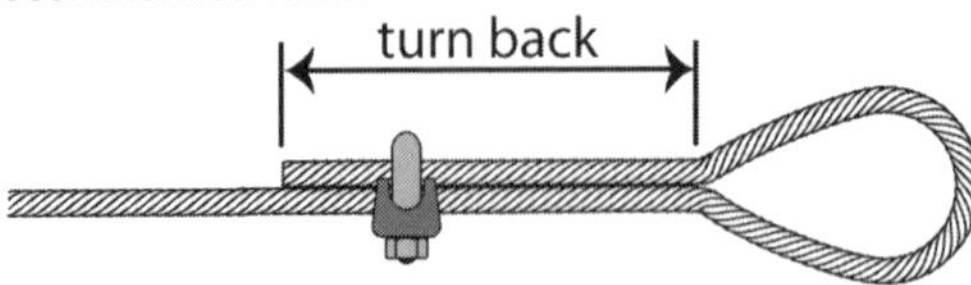
turn back

STEP 2 Apply the second clip as near to the loop as possible. Tighten nuts evenly and torque at the recommended value.

STEP 3 Apply the rest of the clips as indicated in the tables. Space clips accordingly and torque.

Apply light tension and tighten all nuts evenly to specified torque

STEP 4 Recheck and re-tighten nuts after the initial load. This load should be at least equal to loads expected in general use.

Wire rope will stretch slightly causing a reduction in diameter which will slacken the clips. Nuts should be checked at frequent intervals for proper tightness.

U-bolt type

Wire rope clips

Consult the equipment manufacturer for specific installation requirements.

Wire Rope Clips - Installation data

SIZE	Number of clips (minimum)	Turnback	Torque in Foot-lbs
1/8"	2	3-1/4"	4.5
3/16"	2	3-3/4"	7.5
1/4"	2	4-3/4"	15
5/16"	2	5-1/4"	30
3/8"	2	6-1/2"	45
7/16"	2	7"	65
1/2"	3	11-1/2"	65
9/16"	3	12"	95
5/8"	3	12"	95
3/4"	4	18"	130
7/8"	4	19"	225
1"	5	26"	225
1-1/8"	6	34"	225
1-1/4"	7	44"	360
1-3/8"	7	44"	360
1-1/2"	8	54"	360
1-5/8"	8	58"	430
1-3/4"	8	61"	590
2"	8	71"	750
2-1/4"	8	73"	750

Double Saddle

Wire rope clips

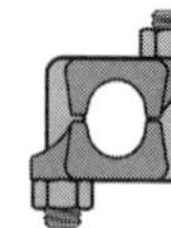

Consult the equipment manufacturer for specific installation requirements.

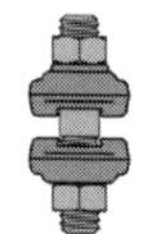

Double Saddle, Wire Rope Clips - Installation data			
SIZE	Number of clips (minimum)	**Turnback**	**Torque in Foot-lbs**
3/16"	2	4"	30
1/4"	2	4"	30
5/16"	2	5"	30
3/8"	2	5-1/4"	45
7/16"	2	6-1/2"	65
1/2"	3	11"	65
9/16"	3	12-3/4"	130
5/8"	3	13-1/2"	130
3/4"	3	16"	225
7/8"	4	26"	225
1"	5	37"	225
1-1/8"	5	41"	360
1-1/4"	6	55"	360
1-3/8"	6	62"	500
1-1/2"	7	78"	500

Crosby® Shackles

Screw Pin Anchor

SIZE

Dia.

Source: Crosby®
G-209 CARBON STEEL Shackles
G-209A ALLOY STEEL Shackles

SIZE	pin dia.	"A" dim	WLL TONS* Carbon	WLL TONS* Alloy
3/16"	0.25	0.38	1/3	-
1/4"	0.31	0.47	1/2	-
5/16"	0.38	0.53	3/4	-
3/8"	0.44"	0.66"	1	2
7/16"	0.5"	0.75"	1 1/2	2 2/3
1/2"	0.63"	0.81"	2	3 1/3
5/8"	0.75"	1.06"	3 1/4	5
3/4"	0.88"	1.25"	4 3/4	7
7/8"	1.00"	1.44"	6 1/2	9 1/2
1"	1.13"	1.69"	8 1/2	12 1/2
1-1/8"	1.25"	1.81"	9 1/2	15
1-1/4"	1.38"	2.03"	12	18
1-3/8"	1.5"	2.25"	13 1/2	21
1-1/2"	1.63"	2.38"	17	-
1-3/4"	2"	2.88"	25	-
2"	2.25"	3.25"	35	-
2-1/2"	2.75"	4.13"	55	-

** Crosby® shackles are rated in METRIC TONS*

Always select and use the shackle by the rated capacity (WLL) that is shown on the shackle. These tables are for reference only - NEVER assume that all shackles are rated the same!

Shackle Side Load

MULTIPLE SLINGS

in shackle body

MAXIMUM included angle

120°

"Multiple slings in the body of a shackle shall not exceed 120 degrees included angle"
ASME B30.26-1.9.4(k)

If the shackle is to be side loaded, the rated load shall be reduced according to the recommendations of the manufacturer or a qualified person per ASME B30.26

SINGLE SLING

in shackle body

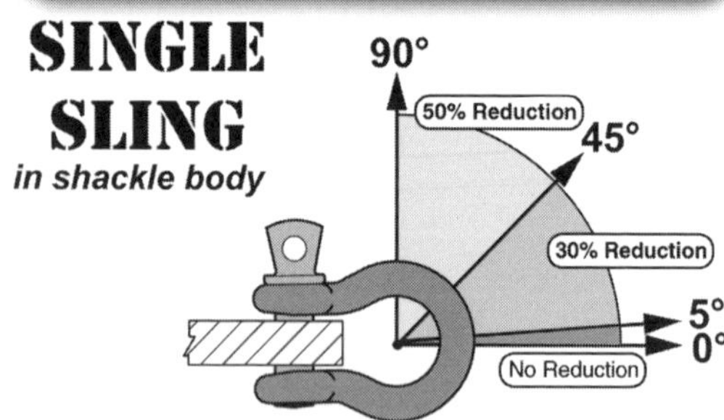
90°
50% Reduction
45°
30% Reduction
5°
0°
No Reduction

	0° to 5° 100%	6° to 45° 30%	46° to 90° 50%
1/2 ton	1,000	700	500
1 ton	2,000	1,400	1,000
2 ton	4,000	2,800	2,000
3 ton	6,000	4,200	3,000
5 ton	10,000	7,000	5,000
10 ton	20,000	14,000	10,000

Generic reductions shown in lbs

NOT APPLICABLE TO ROUND PIN SHACKLES

Web Sling Shackles

Crosby® S-253
Web Sling Shackle

Web Size	Round sling Size	WLL (tons)	A	D	E
2"	# 2	3 1/2	1.06	0.75	1.62
3"	# 3	4 2/4	1.25	0.88	1.5
4"	# 4	6 1/2	1.44	1	2
6"	# 6	8 1/2	1.69	1.13	2.75

Crosby® S-281
Web Sling Shackle

Web Size	Round sling Size	WLL (tons)	A	D	E
1'"	# 1 & 2	3 1/2	0.88	0.75	1.5
1-1/2"	# 3 & 4	6 2/4	1.25	0.88	1.88
2"	# 5 & 6	8 3/4	1.38	1	2.81
3"	# 7 & 8	12 1/2	1.62	1.25	3.06
4"	# 9 & 10	20 1/2	2.12	1.5	5.75
5"	# 11 & 12	35	2.5	2	6.34
6"	# 13	50	3	2.25	7.7

Wide Body Shackles

Crosby® Wide Body Shackles G-2160

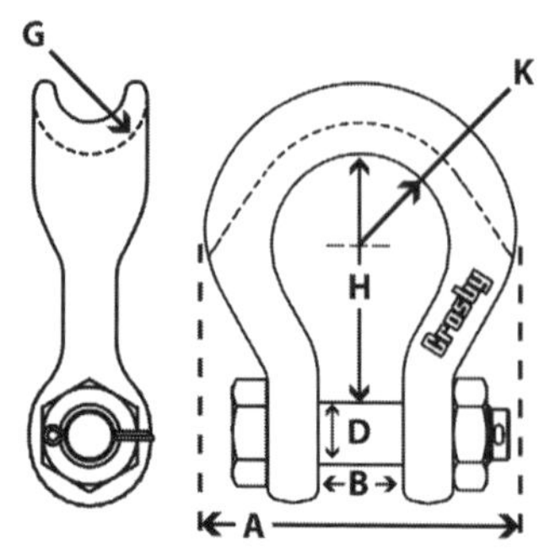
G
K
H
D
B
A
Crosby

WLL*	Shackle Weight *in pounds*
30	25
40	35
55	71
75	99
125	161
200	500
300	811
400	1,041
500	1,378
600	1,833
700	2,446
800	3,016
900	3,436
1,000	4,022

Dimensions are in inches

WLL*	A	B	D	H	K	G
30	7.73	2.37	1.63	7.00	2.50	2.50
40	9.32	2.88	2.00	8.13	3.00	1.75
55	10.41	3.25	2.27	9.42	3.50	2.00
75	14.37	4.13	2.75	11.60	3.64	2.55
125	16.51	5.12	3.15	14.43	4.33	3.15
200	20.67	5.91	4.12	18.98	5.41	4.33
300	24.20	7.38	5.25	23.69	6.31	5.47
400	30.06	8.66	6.30	22.71	7.28	6.30
500	32.99	9.84	7.09	24.88	8.86	6.69
600	35.39	10.83	7.87	27.64	9.74	7.28
700	38.91	11.81	8.46	29.04	10.63	7.87
800	43.50	12.80	9.06	29.62	10.92	8.27
900	43.60	13.78	9.84	30.02	11.51	8.66
1000	45.98	14.96	10.63	30.02	12.11	9.06

*** Working Load Limit (metric tons)**

Slingmax® Twin-Path®
Shackle Sizes

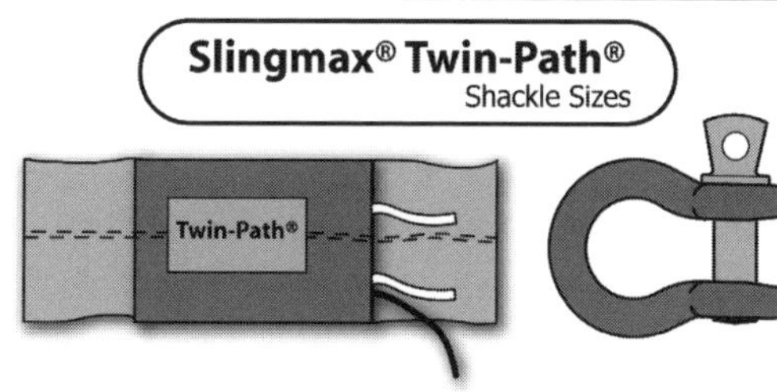
Twin-Path®

WLL's are based on a MINIMUM straight pin diameter, one-half the width of the sling

From the Slingmax® Tech Bulletin #3 "Pin Sizes for Twin-Path® Extra Slings" (December 2003)

Twin-Path® Stock No.	Vertical Rated Capacity	Approx. Body Width (Inches)
TP 200	2,000	2"
TP 300	3,000	2"
TP 450	4,500	2"
TP 600	6,000	3"
TP 750	7,500	3"
TP 900	9,000	3"
TP 1200	12,000	4"
TP 1400	14,000	4"
TP 1700	17,000	4"
TP 2200	22,000	5"
TP 2600	26,000	5"
TP 3200	32,000	5"
TP 5000	50,000	6"
TP 6000	60,000	6"

Shackles must have the same or greater rated capacity as the sling being used. The shackle openings must be of the proper shape and size to assure that the sling will seat properly without bunching.

Twin-Path® information provided by:

SLINGMAX® Rigging Solutions (www.slingmax.com)
Telephone: 610-485-8500

Slingmax® Twin-Path®
Shackle Sizes

Suggested shackles based on a straight pin diameter that is one-half the sling width.

Sling Width	SHACKLE SIZE, TYPE, WLL	PIN DIA	STOCK DIA
2"	7/8" Screw Pin Anchor (9.5 t)	1.000	0.880
	1" Screw Pin Anchor (12.5 t)	1.130	1.000
3"	1-1/2" Bolt Type (30 t)	1.660	1.530
	1-3/4" Bolt Type (40 t)	2.040	1.840
	30 ton Wide Body Shackle	1.630	2.500
4"	1-3/4" Bolt Type (40 t)	2.040	1.840
	2" Bolt Type (55 t)	2.300	2.080
	40 ton Wide Body Shackle	2.000	1.750
	55 ton Wide Body Shackle	2.270	2.000
5"	2-1/2" Bolt Type (85 t)	2.710	2.710
	75 ton Wide Body Shackle	2.750	2.550
6"	3" Bolt Type (120 t)	3.300	3.120
	125 ton Wide Body Shackle	3.150	3.150

Note: It's a good practice to always use as large of a diameter as possible. Since shackle pins are usually larger than the body diameter, selecting the shackle based on the body or stock diameter will result in a larger load bearing surface area for the sling.

When using a shackle that has exposed threads on the pin, cover or protect this area to prevent the sling from being forced into this area when loaded.

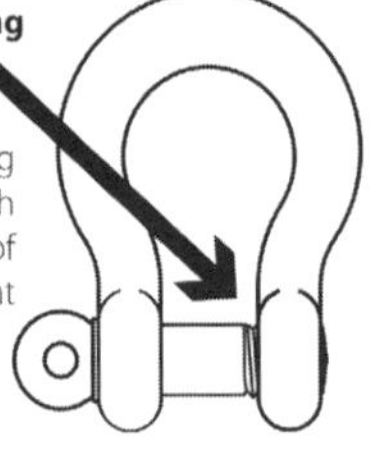

All slings are subject to cutting when lifting on load edges. All edges in contact with the sling must be padded with material of sufficient thickness or strength to prevent damage to the sling.

Alloy Master Link

Crosby A-342

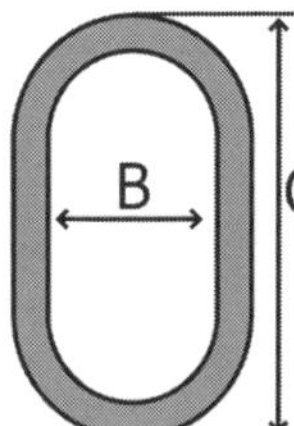
B
C

DIA	WLL	"B"	"C"
1/2"	7,000	2.5	5
5/8"	9,000	3	6
3/4"	12,300	2.75	5.5
7/8"	15,000	3.75	6.38
1"	24,360	3.5	7
1-1/4"	36,200	4.38	8.75
1-1/2"	54,300	5.25	10.5
1-3/4"	84,900	6	12
2"	102,600	7	14
2-1/4"	143,100	8	16

WLL Data taken from Crosby 2006 catalog

120°

Based on Single leg slings (in-line load) or resultant load on multiple legs with an included angle less than or equal to 120 degrees.

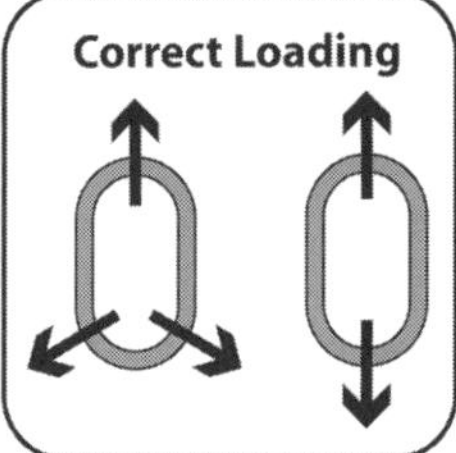
Correct Loading

Improper Loading
NO!

The capacity or WLL is NOT the same for all equipment. The data shown above is only valid when using A-342 Alloy Master Links manufactured by The Crosby Group.

Links by other manufacturers will have different, and usually lower, Working Load Limits.

DO NOT USE THE VALUES SHOWN ABOVE WITH UNKNOWN MANUFACTURERS EQUIPMENT!

Alloy Pear Link

Crosby A-341

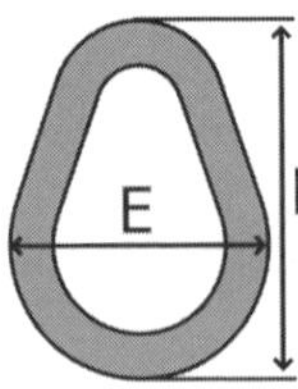
E
D

DIA	WLL	"D"	"E"
1/2"	7,000	2.5	5
5/8"	9,000	3	6
3/4"	12,300	2.75	5.5
7/8"	14,000	3.75	6.38
1"	24,360	3.5	7
1-1/4"	36,000	4.38	8.75
1-1/2"	54,300	5.25	10.5
1-3/4"	84,900	6	12
2"	102,600	7	14
2-1/4"	143,100	8	16

The data shown above is only valid when using A-341 Alloy Pear Shaped Links manufactured by The Crosby Group. WLL Data taken from Crosby 2006 catalog

120°

Based on Single leg slings (in-line load) or resultant load on multiple legs with an included angle less than or equal to 120 degrees.

Improper Loading
NO!

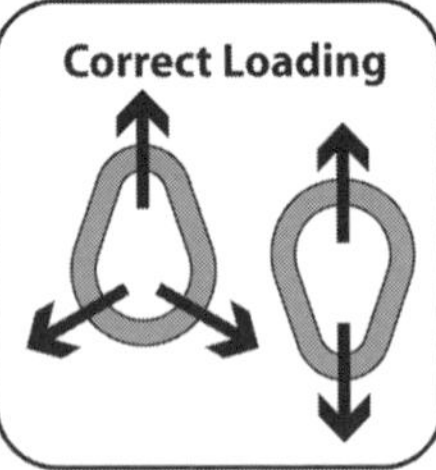
Correct Loading

DO NOT USE THE VALUES SHOWN ABOVE WITH UNKNOWN MANUFACTURERS EQUIPMENT!

Turnbuckles

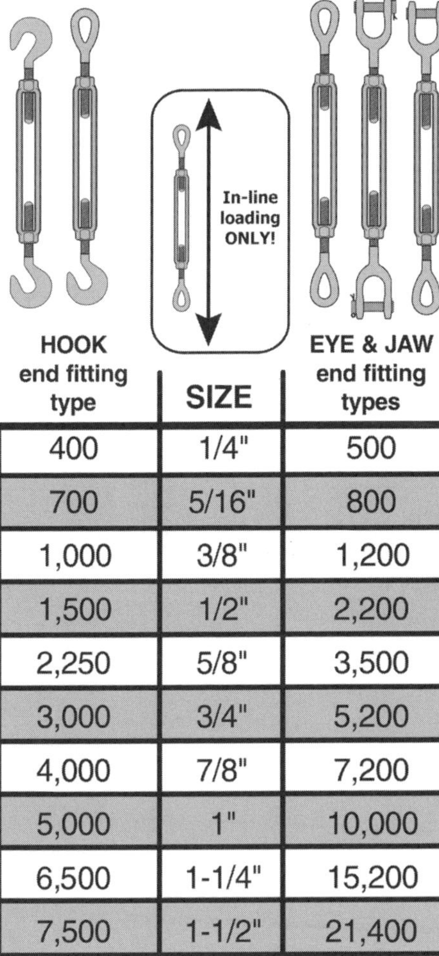
In-line
loading
ONLY!

HOOK end fitting type	SIZE	EYE & JAW end fitting types
400	1/4"	500
700	5/16"	800
1,000	3/8"	1,200
1,500	1/2"	2,200
2,250	5/8"	3,500
3,000	3/4"	5,200
4,000	7/8"	7,200
5,000	1"	10,000
6,500	1-1/4"	15,200
7,500	1-1/2"	21,400

WLL values shown in pounds

Weldless Rings

Crosby S-643

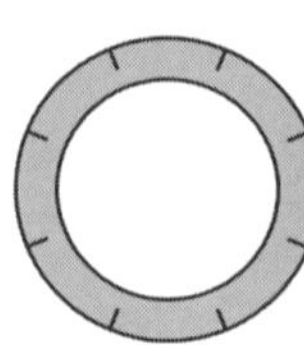

DIA	WLL
7/8" Dia X 4" ID	7,200
7/8" Dia X 5-1/2" ID	5,600
1" Dia X 4" ID	10,800
1-1/8" Dia X 6" ID	10,400
1-1/4" Dia X 5" ID	17,000
1-3/8" Dia X 6" ID	19,000

The data shown above is only valid when using Crosby S-643 Weldless Rings manufactured by The Crosby Group. WLL Data taken from Crosby 2006 catalog

DO NOT USE THE VALUES SHOWN ABOVE WITH UNKNOWN MANUFACTURERS EQUIPMENT!

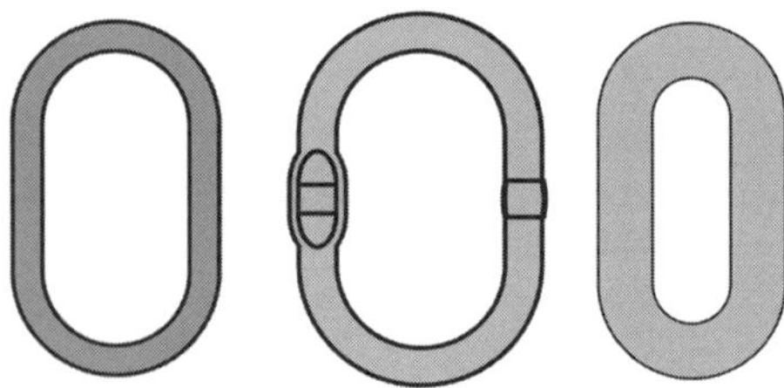

Just because they may look similar DOES NOT mean they are the same!

Use caution when selecting and using any link or ring. These useful pieces of rigging hardware vary widely between manufacturers. Some master links are made for use with chains and have a different WLL than a standard master link. Components manufactured for use with chains may only have a 4:1 Design Factor instead of the 5:1 required for other types of slings and rigging hardware.

SLINGS
HARDWARE
FORMULAS
EXAMPLES
REFERENCE
STANDARDS
EXTRAS

FORMULAS

OSHA 1910.184 states that "slings shall not be used with loads in excess of their rated capacities".

The weight of the object to be lifted must be established before you can select the rigging and rigging hardware.

Standard weights of typical materials

Material	Cu. ft	Cu. Inch
Aluminum	165.00	0.0955
Brass	535.00	0.3096
Brick masonry, common	125.00	0.0723
Bronze	500.00	0.2894
Cast Iron	480.00	0.2778
Cement, portland, loose	94.00	0.0544
Concrete, stone aggr.	144.00	0.0833
Copper	560.00	0.3241
Earth, dry	75.00	0.0434
Earth, wet	100.00	0.0579
Glass	160.00	0.0926
Ice	56.00	0.0324
Lead	710.00	0.4109
Snow, fresh fallen	8.00	0.0046
Snow, wet	35.00	0.0203
Steel	490.00	0.2836
Tin	460.00	0.2662
Water	62.00	0.0359
Gypsum wall board	54.00	0.0313
Wood, pine	30.00	0.0174

Cube or rectangle

1. Determine the volume of the object
   a × b × c = volume

2. Determine the approximate weight of the object
   volume × weight (per Cu)

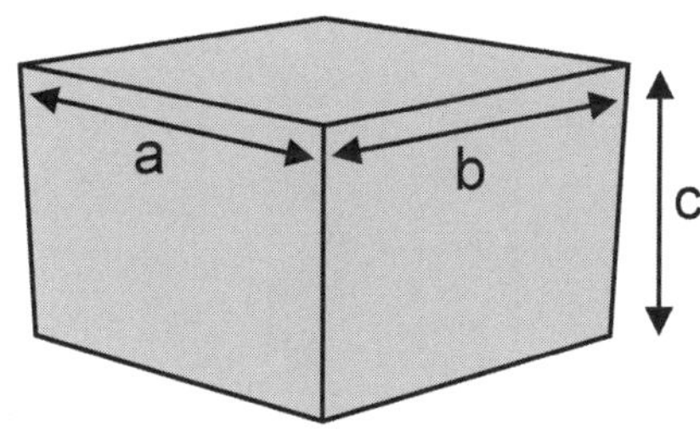

a
b
c

EXAMPLE:

a = 12 inches **c** = 5 inches
b = 24inches **Material:** Steel

Step 1 **find the volume:**
12 X 24 X 5 = 1440 Cu inches

Step 2 **find the weight:**
1440 X .2836 = 408.38 lb

Answer: **408 pounds**

SLINGS HARDWARE FORMULAS EXAMPLES REFERENCE STANDARDS EXTRAS

1. Determine the volume of the object
   VOLUME = 0.7854 X D X D X H

2. Determine the approximate weight of the object
   WEIGHT = Volume X Weight (per Cu)

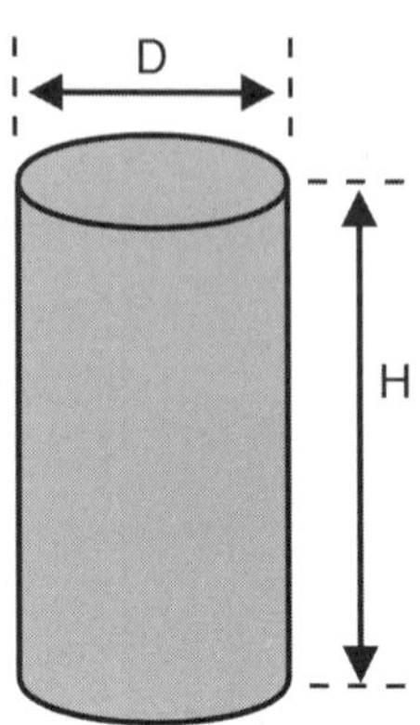
D
H

EXAMPLE:

D = 18 inches
H = 42inches **Material:** Steel

Step 1 **find the volume:**

0.7854 X 18 X 18 X 42 = 10687.72 Cu in

Step 2 **find the weight:**

10687.72 X .2836 = 3031.03 lbs

Answer: **3031 pounds**

Round & Hollow (Pipe)

1. Determine the volume of the object
   VOLUME = T × (D-T) × 3.141 × H

2. Determine the approximate weight of the object
   WEIGHT = Volume × Weight (per Cu)

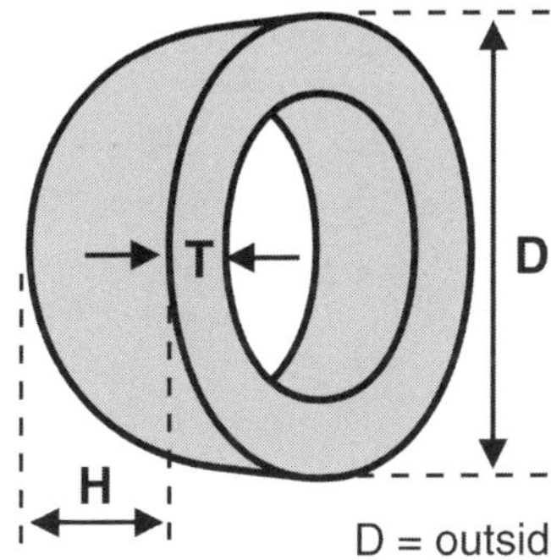
T
D
H

D = outside diameter
H = overall length of pipe
T = wall thickness

EXAMPLE:

D = 18 inches **Material:** Steel
T = 1.5 inches
H = 48 inches

Step 1 **find the volume:**

1.5 × (18 -1.5) × 3.141 × 48 = 3731.5 Cu in

Step 2 **find the weight:**

3731.5 × .2836 = 1058.25 lbs

Answer: **1058 pounds**

Frustum of a cone

1 Determine the volume of the object

VOLUME = 0.2618 X h X (D^2 + Dxd + d^2)

2 Determine the approximate weight of the object

WEIGHT = Volume × Weight (per Cu)

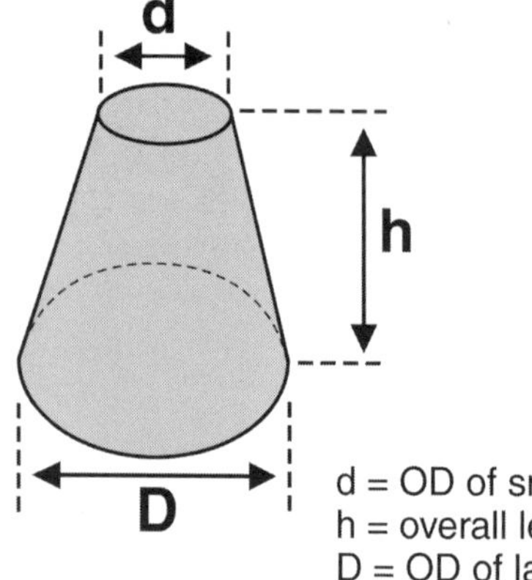
d
h
D

d = OD of small end
h = overall length
D = OD of large end

EXAMPLE:

d = 6 inches
D = 18 inches
h = 32 inches

Material: Steel

Step 1 find the volume:

0.2618 × 32 × ((18×18)+(18×6)+(6×6))= 3920.71 Cu in

also expressed as: 0.2618 × 32 × (468) = 3920.71 Cu in

Step 2 find the weight:

3920.71 × .2836 = 1111.91 lbs

Answer: 1112 pounds

The four shapes covered on the previous pages will allow you to determine the approximate weight of almost any object. For example examine the illustration of a valve bonnet below, notice that if you visualize it as separate shapes you can easily determine its approximate weight.

When in doubt, estimate on the HIGH side. It is always better to OVER RIG than UNDER RIG.

The following applies to the illustration of the valve bonnet shown below.

1.) Is considered round & hollow.

2.) Is a frustum of a cone.

3.) Is the hole in the center.

(figure the hole as a solid round shape, then subtract its weight from the fustum.)

4.) Is considered round & hollow.

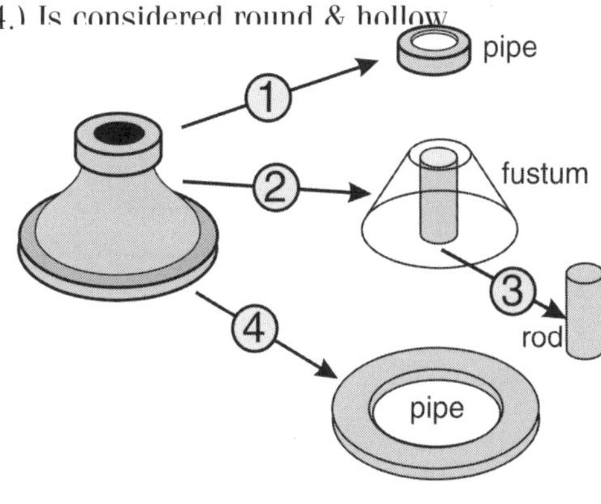
1
pipe
2
fustum
3
rod
4
pipe

Remember, establishing the weight of an object is one of the most important steps in determining what type of rigging equipment to use for a safe lift!

SLINGS HARDWARE FORMULAS EXAMPLES REFERENCE STANDARDS EXTRAS

Tension Basics
Sling Angles

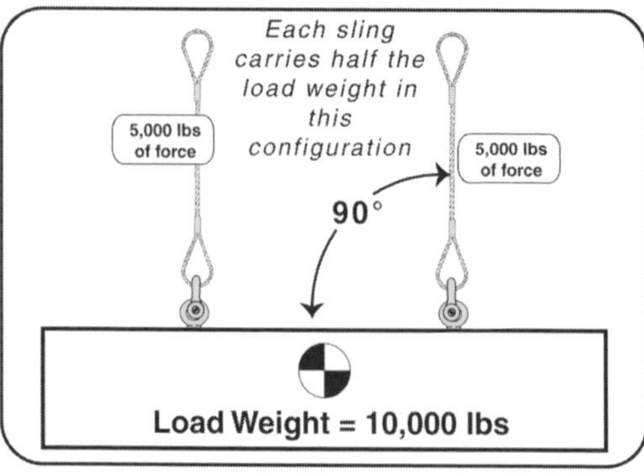
Each sling carries half the load weight in this configuration
5,000 lbs of force
5,000 lbs of force
90°
Load Weight = 10,000 lbs

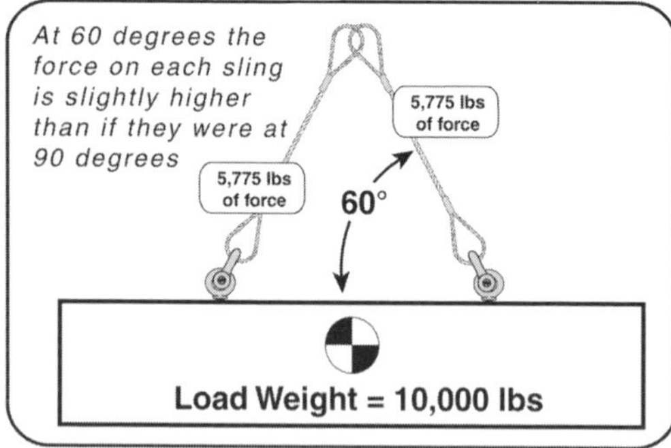
At 60 degrees the force on each sling is slightly higher than if they were at 90 degrees
5,775 lbs of force
5,775 lbs of force
60°
Load Weight = 10,000 lbs

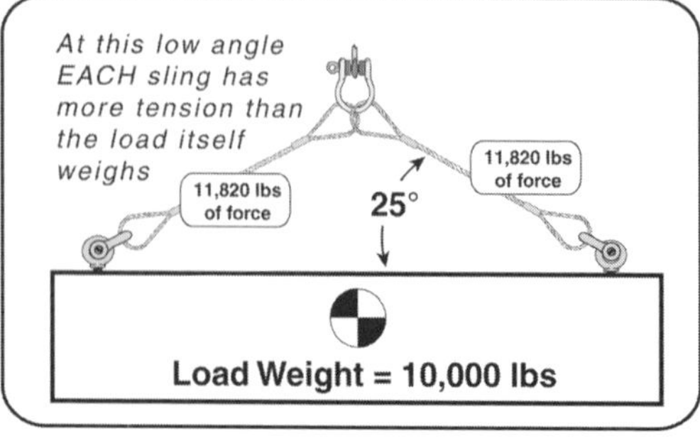
At this low angle EACH sling has more tension than the load itself weighs
11,820 lbs of force
11,820 lbs of force
25°
Load Weight = 10,000 lbs

Quick Check

Sling Angles

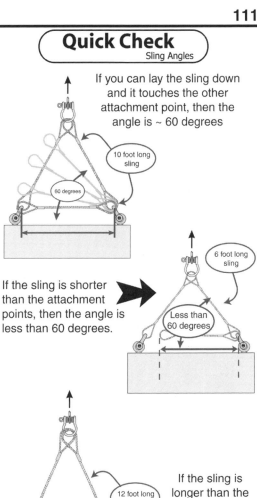

If you can lay the sling down and it touches the other attachment point, then the angle is ~ 60 degrees

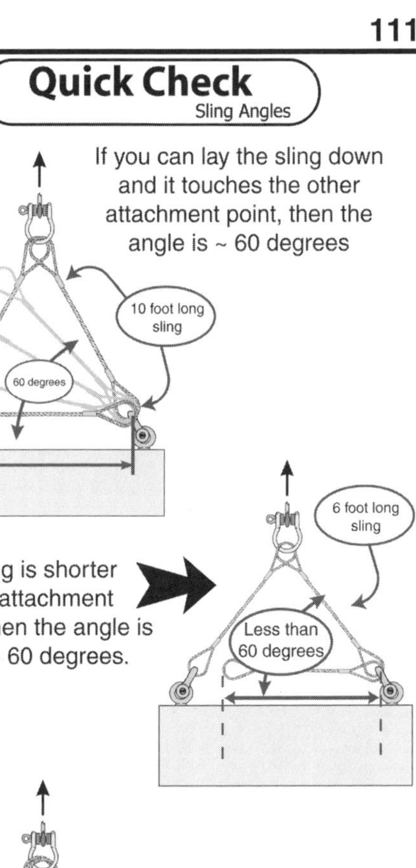
10 foot long sling
60 degrees
6 foot long sling
Less than 60 degrees

If the sling is shorter than the attachment points, then the angle is less than 60 degrees.

12 foot long sling
More than 60 degrees

If the sling is longer than the attachment points, then the angle is greater than 60 degrees.

Leg Loading

Graphics ©2007 Jerry Klinke

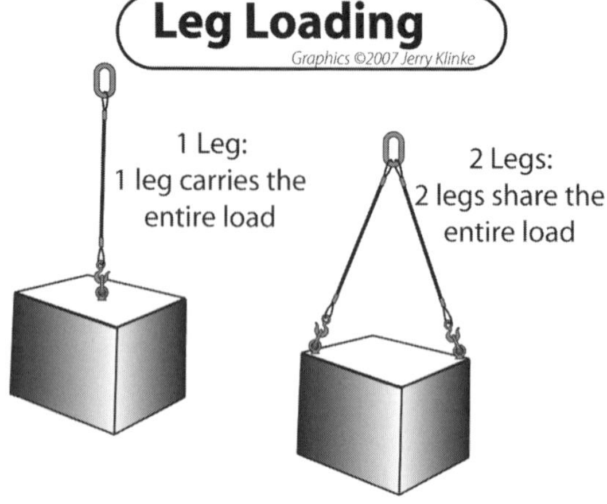

1 Leg:
1 leg carries the entire load

2 Legs:
2 legs share the entire load

It's recommended you consider that only 2 legs will carry the load, even when using 4 legs, since it is difficult to assure that all the legs will carry an equal share of the load.

3 Legs:
3 legs share the entire load

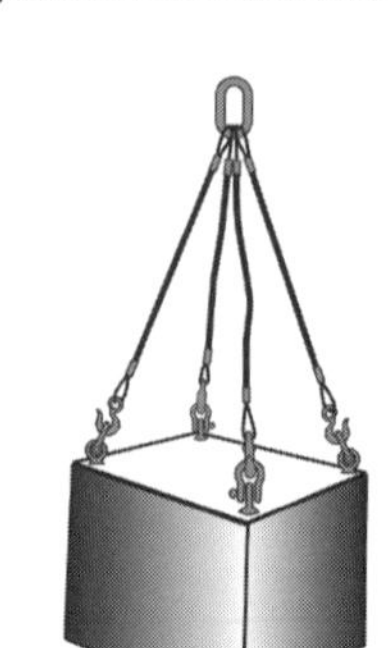

4 Legs:
2 legs carry a majority of the load, the other 2 legs help balance the load

Angular Tension

Graphics ©2007 Jerry Klinke

Remember:

"low angles ALWAYS increase the tension"

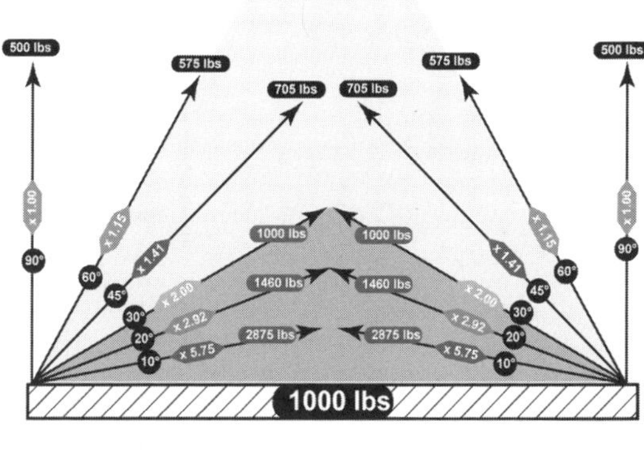
500 lbs
575 lbs
705 lbs
705 lbs
575 lbs
500 lbs
x 1.00
x 1.15
x 1.41
1000 lbs
1000 lbs
1460 lbs
1460 lbs
2875 lbs
2875 lbs
x 2.00
x 2.92
x 5.75
90°
60°
45°
30°
20°
10°
1000 lbs

It's best to try to maintain a 60 degree horizontal sling angle if possible

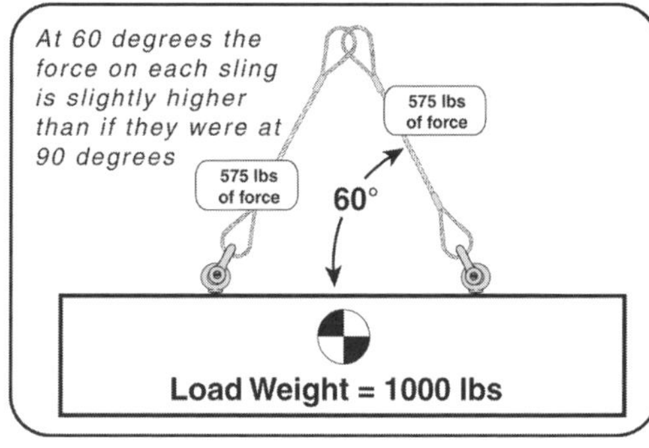
At 60 degrees the force on each sling is slightly higher than if they were at 90 degrees
575 lbs of force
575 lbs of force
60°
Load Weight = 1000 lbs

Tension Calculation

Using Horizontal Sling Angle

To determine the amount of tension on a sling used at angles other than 90 degrees (vertical), use the table at the right to obtain the Load Angle Factor (L.A.F.) and following formula:

(Weight ÷ No of legs) x L.A.F.

Horizontal Angle	L.A.F.
5 °	11.49
10 °	5.75
15 °	3.861
20 °	2.924
25 °	2.364
30 °	2.00
35 °	1.742
40 °	1.555
45 °	1.414
50 °	1.305
55 °	1.221
60 °	1.155
65 °	1.104
70 °	1.064
75 °	1.035
80 °	1.015
85 °	1.004
90 °	1.00

Example: If the load weight is 4,000 lbs, and two (2) slings are used at a 40 degree angle each.

(4000 ÷2) x 1.555 = 3,110

Therefore, each leg will have 3,110 lbs of tension.

It is recommended that you consider only 2 legs will carry the load, even when using 3 and 4 legs, since it is difficult to assure that all legs will carry an equal share of the load.

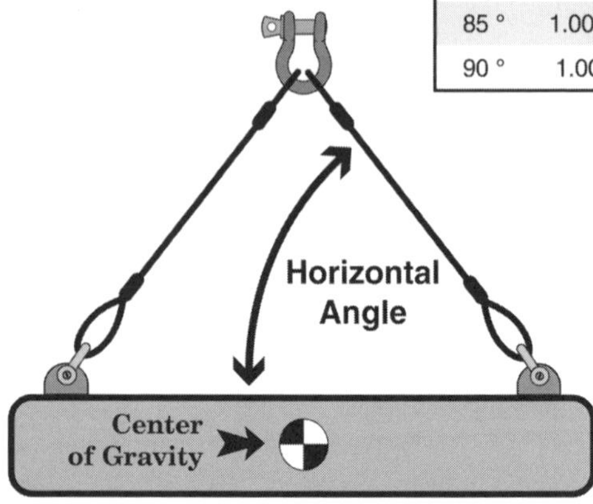
Horizontal Angle
Center of Gravity

When the load is NOT distributed equally on all slings, the tension on each leg must be calculated individually by a qualified person.

Angle Finder

On the back of this book

You can determine the horizontal angle by using the Angle Finder that is on the back of this book. All you need is a short piece of string, a nut or washer and your thumb!

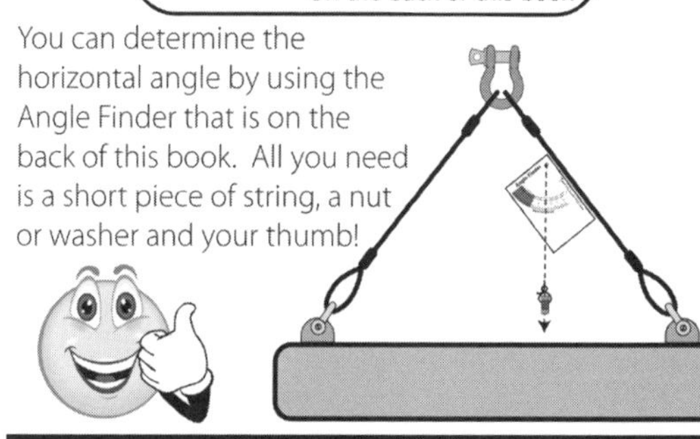

1) Attach a weight (nut or washer) to a short piece of string.
2) Place the edge of the book against the sling.
3) Hold the string in place with your thumb and forefinger.
4) Let the string hang freely and read the angle on the scale.

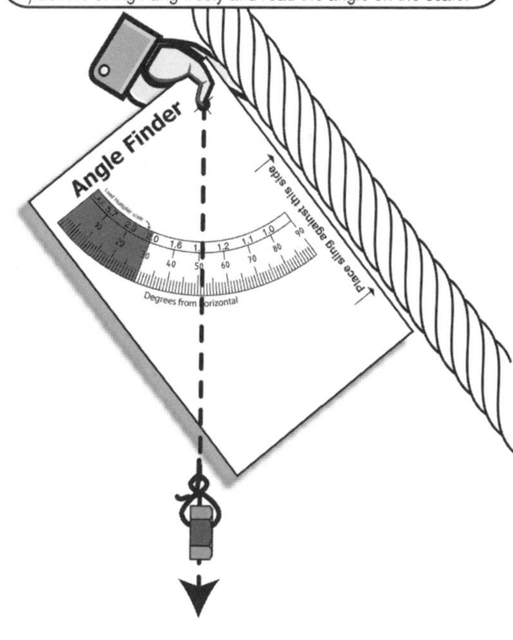
Angle Finder
Place sling against this side
1.6
1.2
1.1
1.0
40
50
60
70
80
90

SLINGS HARDWARE FORMULAS EXAMPLES REFERENCE STANDARDS EXTRAS

Tension Calculation
Using measurements

It's hard to determine the exact angle when working in the field unless you have a protractor handy. The following formula provides accurate calculations by using only measurements taken in the field:

(Weight ÷ No of legs) X (S ÷ H)

Example: The load weight is 6,000 lbs and two (2) slings are used. You measure up the sling 36" (this is the "S" dimension) then measure straight down and obtain a 24" measurment (this is the "H" dimension).

weight = 6,000 lbs
"S" = 36 inches
"H" = 24 inches

(6000 ÷ 2) x (36 ÷ 24)

3000 x 1.5 = 4,500 lbs of tension per leg

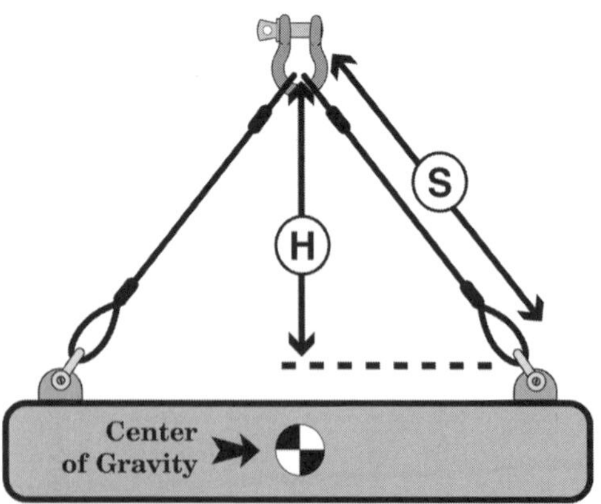
S
H
Center of Gravity

It is recommended that you consider that only 2 legs will carry the load, even when using 3 and 4 legs, since it is difficult to assure that all legs will carry an equal share of the load.

"For multiple-leg slings used with nonsymmetrical loads, an analysis by a qualified person should be performed to prevent overloading of any leg" ASME B30.9

Tension Calculation

Using measurements

Instead of measuring the sling length, you can also obtain the "S" & "H" dimensions by creating an "invisible triangle" with a tape measure (remember that all measruements must be in INCHES).

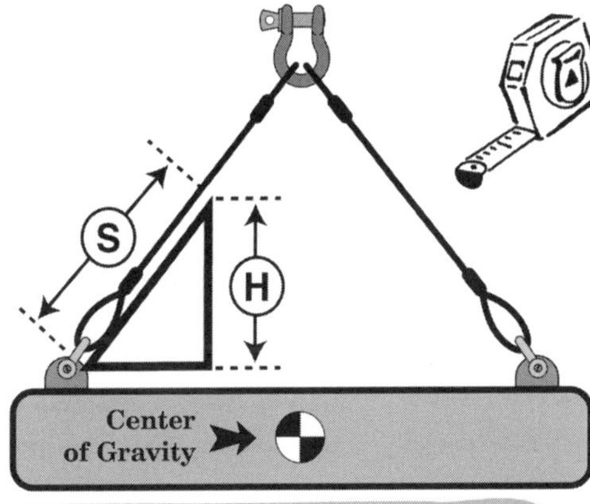
S
H
Center
of Gravity

IMPORTANT: Make sure when measuring the "H" dimension that the tape is perfectly plumb (straight down) otherwise it will not be a true right angle triangle.

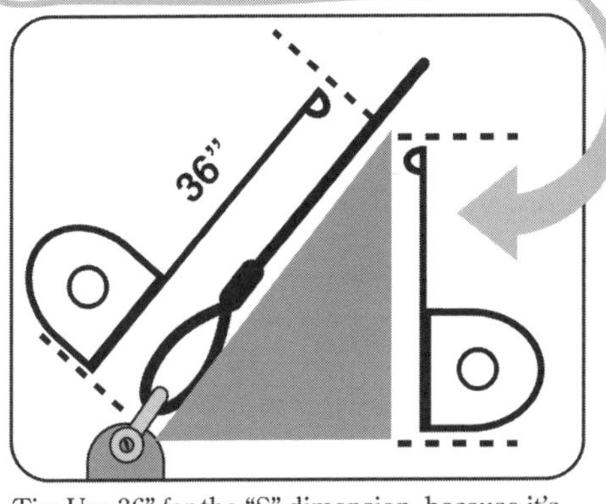
36"

Tip: Use 36" for the "S" dimension, because it's easy to remember.

Drifting Loads
with chainfalls

To determine how much tension will be placed upon chainfalls used in angular rigging situations, use the following formula:

Tension on Chainfall "A" =
(Load weight × D2 × LA) ÷ (H × D3)

Tension on Chainfall "B" =
(Load weight × D1 × LB) ÷ (H × D3)

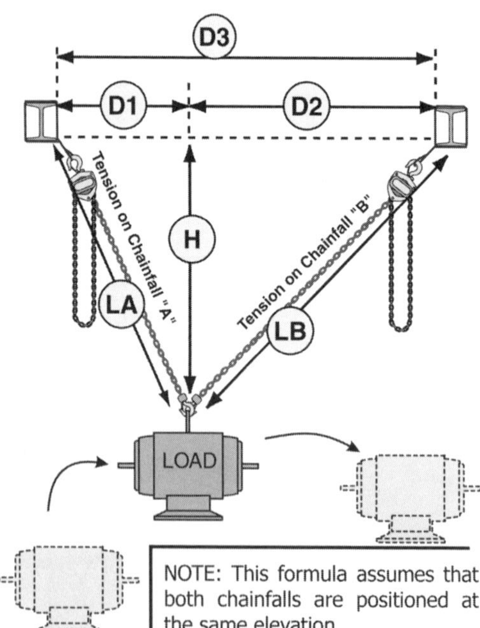
D3
D1
D2
H
LA
LB
Tension on Chainfall "A"
Tension on Chainfall "B"
LOAD

NOTE: This formula assumes that both chainfalls are positioned at the same elevation.

Drifting Loads
EXAMPLE

Values used in this example:

D1 = 48"　D2 = 96"　D3 = 144"
LA = 60"　LB = 102.5"　H = 36"

Weight of load = 2,000 lbs

Chainfall "A"

$$(\text{Load} \times D2 \times LA) \div (H \times D3) = A$$

2000 x 96 x 60 = 11,520,000 ÷ 36 x 144 = 5,184

11,520,000 ÷ 5,184 = 2,222

Answer:

Chainfall "A" will carry 2,222 lbs of tension

Chainfall "B"

$$(\text{Load} \times D1 \times LB) \div (H \times D3) = B$$

2000 x 48 x 102.5 = 9,840,000 ÷ 36 x 144 = 5,184

9,840,000 ÷ 5,184 = 1,898

Answer:

Chainfall "B" will carry 1,898 lbs of tension

Tightlines

To determine how much tension will be placed upon slings used in angular rigging situations, use the following formula:

Tension on the Left Sling (LS) = (Load × D2 × LS) ÷ (H × D3)

Tension on the Right Sling (RS) = (Load × D1 × RS) ÷ (H × D3)

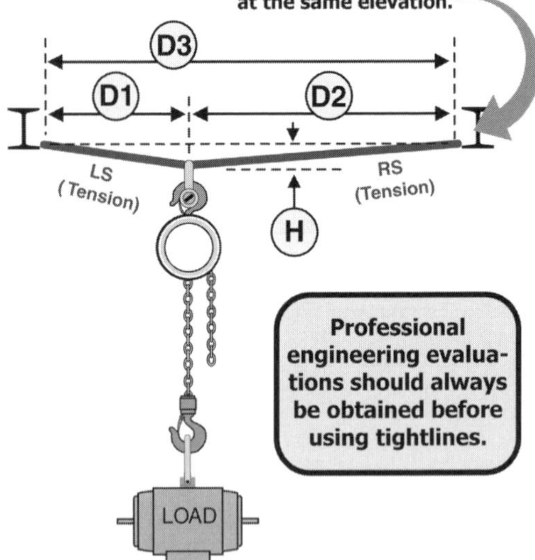
NOTE:
This formula is based on both sides being attached at the same elevation.
D3
D1
D2
LS
(Tension)
RS
(Tension)
H
LOAD
Professional engineering evaluations should always be obtained before using tightlines.

Tightlines can create tremendous forces on structures and rigging equipment. This formula only considers the additional tension placed upon the slings. Considerations must also be made for the shackles, attachments, and building members used to rig from. Professional engineering evaluations should always be obtained when using tightlines.

Tightlines
EXAMPLE

Values used in this example:

D1 = 40" D2 = 105" D3 = 145"
LS = 48" RS = 120" H = 28"

Weight of load = 1,000 lbs

LEFT sling

(Load × D2 × LS) ÷ (H × D3)

1000 x 105 x 48 = 5,040,000 ÷ 28 x 145 = 4,060

5,040,000 ÷ 4,060 = 1,241

Answer:

The LEFT sling will have 1,241 lbs of tension

RIGHT sling

(Load × D1 × RS) ÷ (H × D3)

1000 x 40 x 120 = 4,800,000 ÷ 28 x 145 = 4,060

4,800,000 ÷ 4,060 = 1,182

Answer:

The RIGHT sling will have 1,182 lbs of tension

Tension Calculation

Unequeal Legs

FORMULA

Sling "A" Tension
(Load x D2 x SA) ÷ H x (D1 + D2)

Sling "B" Tension
(Load x D1 x SB) ÷ H x (D1 + D2)

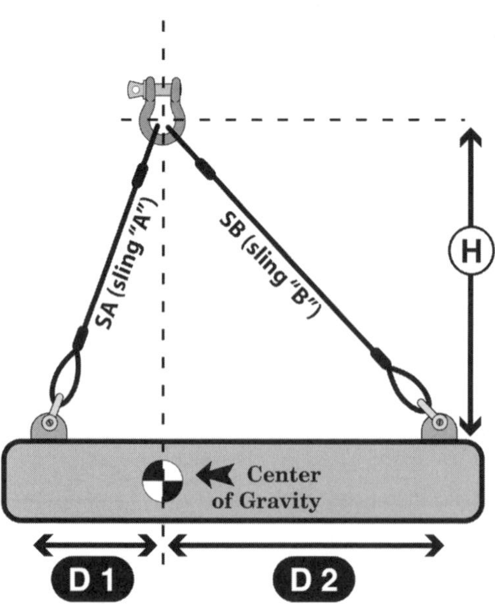
SA (sling "A")
SB (sling "B")
H
Center of Gravity
D 1
D 2

Unequal Legs
EXAMPLE

Values used in this example:

D1 = 40" D2 = 63" H = 60"
SA = 66" SB = 78"

Weight of load = 10,000 lbs

Sling "A"

(Load x D2 x SA) ÷ H x (D1 + D2)

10,000 x 63 x 66 = 41580000 ÷ 60 x (40 + 63) = 6,180

41580000 ÷ 6,180 = 6,728 lbs

Answer:

Sling "A" will have 6,728 lbs of tension

Sling "B"

(Load x D1 x SB) ÷ H x (D1 + D2)

10,000 x 40 x 78 = 31200000 ÷ 60 x (40 + 63) = 6,180

31200000 ÷ 6,180 = 5,049 lbs

Answer:

Sling "A" will have 6,728 lbs of tension

Tension Calculation
Unequeal Legs & Heights

Calculating the stress on non-symmetrical bridle hitches is more complex that the normal sling formula. Because of the imbalance of load weight, you must determine the center of gravity first. Then the sling tensions can be determined using the formulas listed below.

FORMULAS

Sling "A" Tension
(Load x D2 x SA) ÷ [(D2 x H1) + (D1 x H2)]

Sling "B" Tension
(Load x D1 x SB) ÷ [(D2 x H1) + (D1 x H2)]

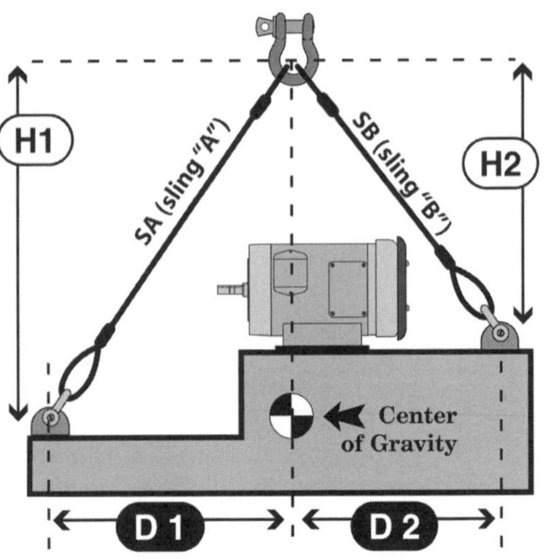
H1
SA (sling "A")
SB (sling "B")
H2
Center
of Gravity
D 1
D 2

When the load is NOT distributed equally on all slings, the tension on each leg must be calculated individually by a qualified person.

Unequeal Legs

EXAMPLE

Values used in this example:

D1 = 55" D2 = 27" H1 = 60"
SA = 81" SB = 36" H2 = 24"

Weight of load = 6,000 lbs

Sling "A"

(Load × D2 × SA) ÷ [(D2 × H1) + (D1 × H2)]

6000 x 27 x 81 = 13122000 ÷ 27 x 60 = 1,620 + 55 x 24 = 1,320

1,620 + 1,320 = 2,940

13122000 ÷ 2,940 = **4,463**

Answer:

Sling "A" will have 4,463 lbs of tension

Sling "B"

(Load × D1 × SB) ÷ [(D2 × H1) + (D1 × H2)]

6000 x 55 x 36 = 11880000 ÷ 27 x 60 = 1,620 + 55 x 24 = 1,320

1,620 + 1,320 = 2,940

11880000 ÷ 2,940 = **4,040**

Answer:

Sling "B" will have 4,040 lbs of tension

Center of Gravity

A simple way to determine the center of gravity (CG) is to weight both ends using a load cell or crane scale. And when you add the weight of the two ends together it will give you the total weight of the load*

*it may be slightly heavier than actual, but very close

Step 1:
Lift each end (about an inch) and record the weights

Step2:
Measure the distance between the lifting/support points

Step 3:
Calculate the CG (From end "A" towards end "B")

$$CG = B \div (A + B) \times D$$

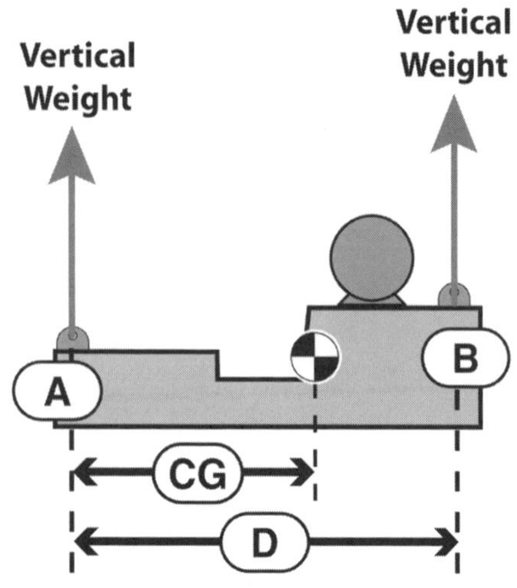
Vertical Weight
Vertical Weight
A
B
CG
D

Center of Gravity

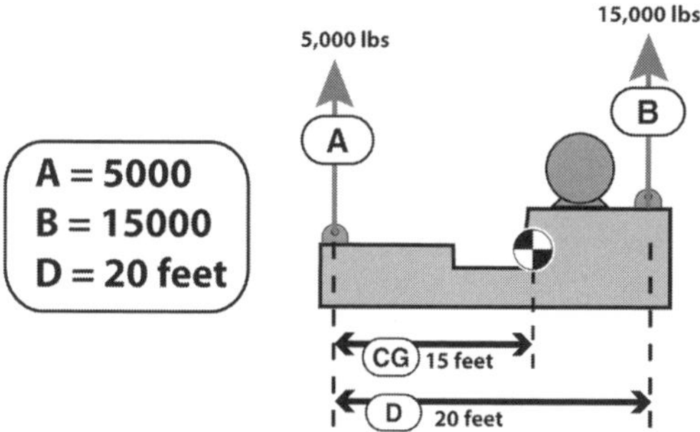
5,000 lbs
15,000 lbs
A
B
A = 5000
B = 15000
D = 20 feet
CG 15 feet
D 20 feet

15000 ÷ (5000 + 15000) X 20 = 15

The CG (from end "A" towards end "B") is 15 ft

This method will provide you with the location of the CG for the side of the load. You can perform the same calculations from the end of the load, then intersect the 2 CG's to find the total of the entire object.

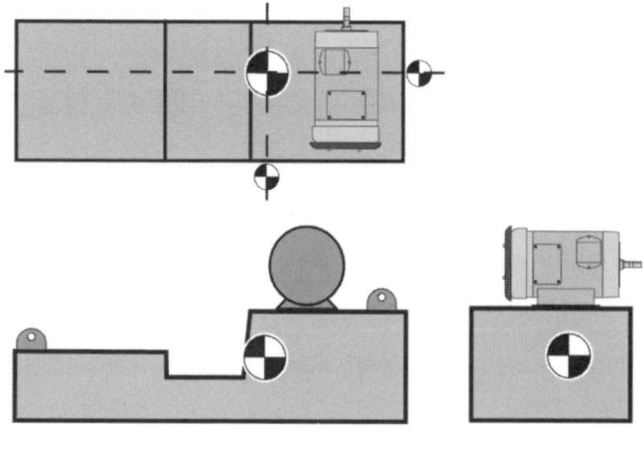

SLINGS HARDWARE FORMULAS EXAMPLES REFERENCE STANDARDS EXTRAS

Block Loading

A single sheave block used to change load line direction can be subjected to total loads greatly different from the weight being lifted or pulled. The total load value varies with the angle between the incoming and departing lines to the block.

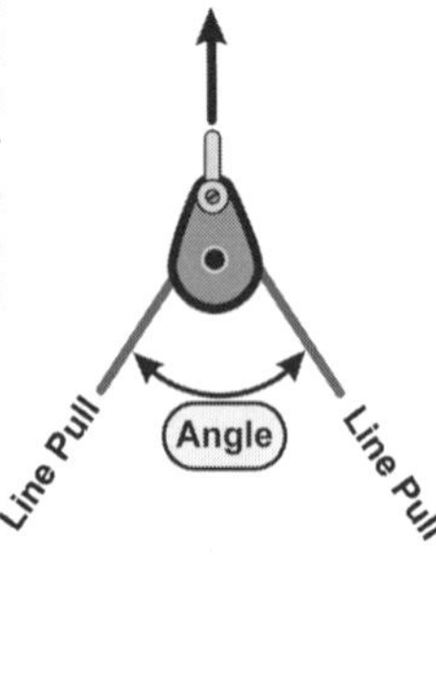
Total Load
Angle
Line Pull
Line Pull

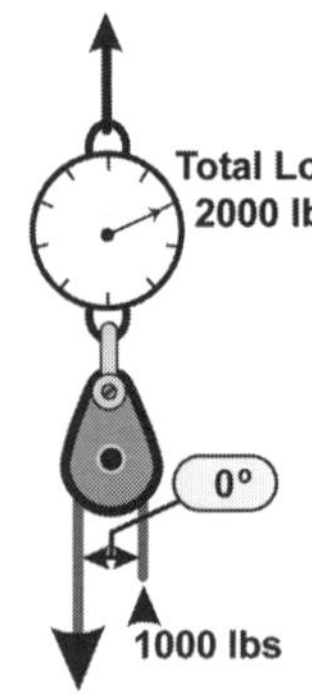
Total Load
2000 lbs
0°
1000 lbs
Line pull
requires
1000 lbs

NOTE: The total load at the pulley attachment point is always greater than the weight of the load being lifted!

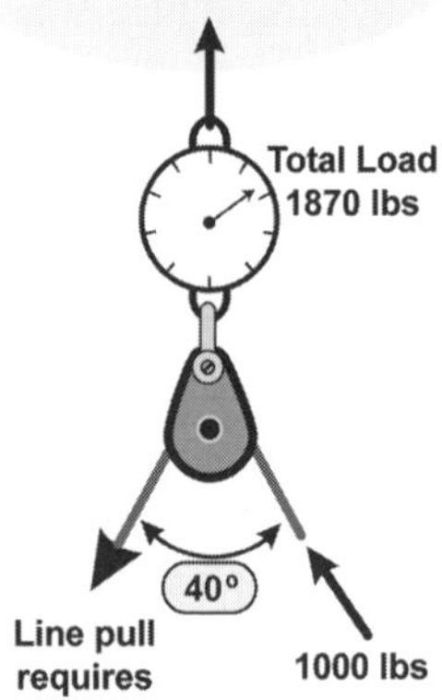
Total Load
1870 lbs
40°
Line pull
requires
1000 lbs
1000 lbs

Block Loading

Total Load = Line Pull x Angle Factor

Example: 5000 lbs line pull at 40° (5000 x 1.87) a total load of 9,350 lbs

Angle	Factor	Angle	Factor
0°	2.00	100°	1.29
10°	1.99	110°	1.15
20°	1.97	120°	1.00
30°	1.93	130°	0.84
40°	1.87	135°	0.76
45°	1.84	140°	0.68
50°	1.81	150°	0.52
60°	1.73	160°	0.35
70°	1.64	170°	0.17
80°	1.53	180°	0
90°	1.41	--	--

Mechanical Advantage

$$\text{Pull Required} = \frac{\text{Weight to be lifted}}{\text{Mechanical Advantage}}$$

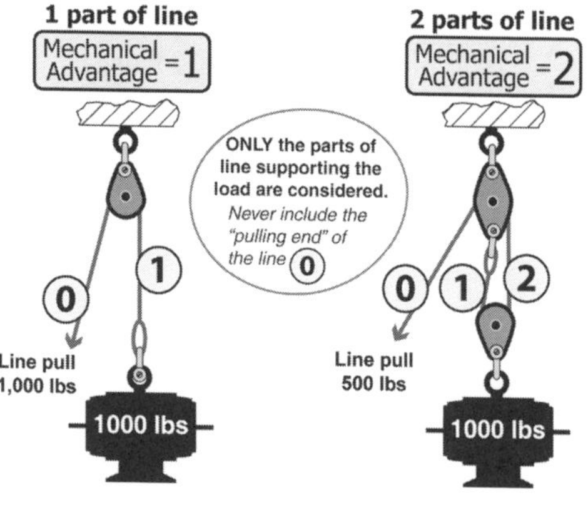

1 part of line
Mechanical Advantage =1
2 parts of line
Mechanical Advantage =2
ONLY the parts of line supporting the load are considered.
Never include the "pulling end" of the line 0
0
1
0
1
2
Line pull 1,000 lbs
Line pull 500 lbs
1000 lbs
1000 lbs

Block Loading
EXAMPLE

Since the block "A" is a traveling block, the mechanical advantage is 2 because two parts of load line support the 1,000 lbs load (Block bearing friction NOT considered).

To determine line pull (force required to lift load):

1000 ÷ 2 = 500 lbs

To determine load on block "A":

(line pull) × (factor of 0 degrees)

500 lbs × 2.00 = 1,000 lbs

To determine load on block "B":

500 lbs × 1.87 + 500 lbs = 1,435 lbs

To determine load on block "C":

500 lbs × .84 = 420 lbs

To determine load on block "D":

500 lbs × 1.41 = 705 lbs

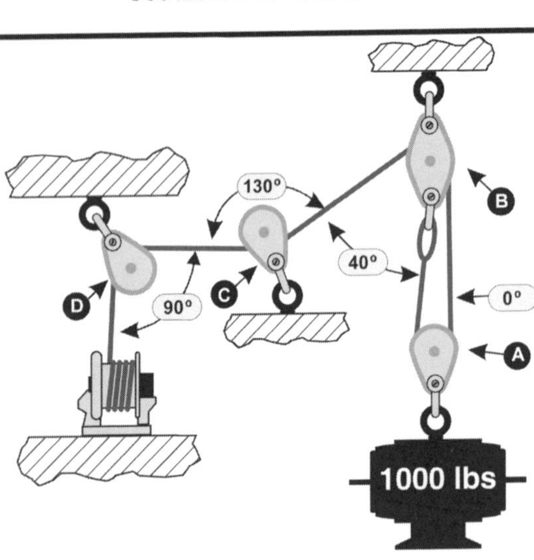
130°
40°
0°
90°
A
B
C
D
1000 lbs

Pulling Force

To move a load on a LEVEL plane

F (Force) = CF x W

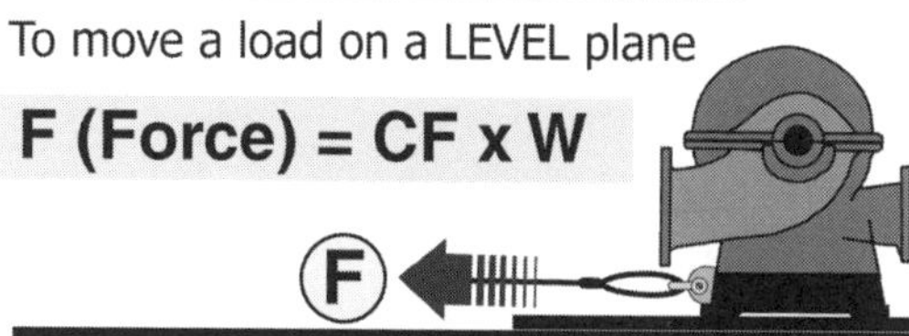
F

To move a load on an UPHILL incline

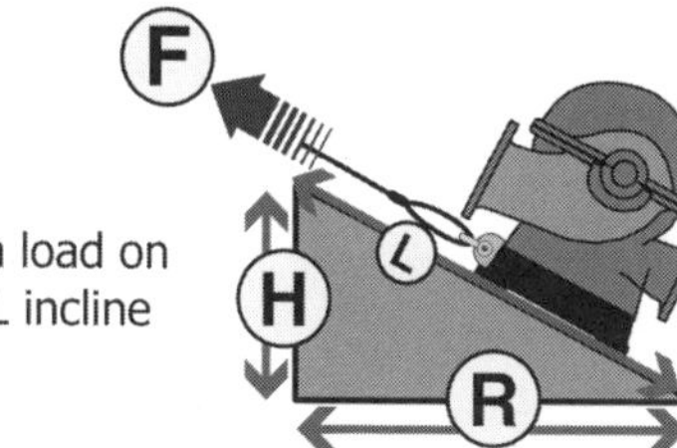
F
L
H
R

F= [CF x W x(R÷L)] + [(H÷L) x W]

To move a load on a DOWNHILL incline:

F= [CF x W x(R÷L)] - [(H÷L) x W]

LEGEND	
F =	Force required to move load
CF =	Coefficient of friction
L =	Length of ramp
H =	HEIGHT *(Vertical distance in feet)*
R =	RUN *(Horizontal distance in feet)*
W =	Weight of load

Surfaces	CF
Concrete on concrete	0.65
Metal on concrete	0.60
Steel on steel	0.20
Cast iron on steel	0.25
Load on wheels	0.05
Load on ice	0.01
Wood on wood	0.50
Wood on metal	0.30
Wood on concrete	0.45
Continuous Lubricated Surface	0.15

These coefficient of friction values apply to hard, clean surfaces sliding against each other. These may not directly relate to your application due to actual surface conditions.

Equipment update

Digital Dynamometers
(sometimes called load cells)
are an invaluable tool to safely and conveniently eliminate guesswork and risks associated with lifting loads of unknown weight.

13,680

They can be used to weight objects, and to measure force on slings or other lifting situations. They are available as a stand alone measuring unit, or they can transmit a wireless signal to a small handheld device that can monitor several units at the same time. Relatively inexpensive compared to typical rigging equipment, these units are also available for rental from many rigging shops and suppliers.

EXAMPLES

Softeners
Graphics ©2007 Jerry Klinke

WEAR PROTECTION

"Sharp edges in contact with the sling should be padded with material of sufficient strength to protect the sling."

ASME B30.9

CORNER PROTECTION

Best practice: change the profile of a corner in contact with a sling to a radius.

Softeners

Graphics ©2007 Jerry Klinke

Sharp corners can cut and damage slings and result in sling failure.

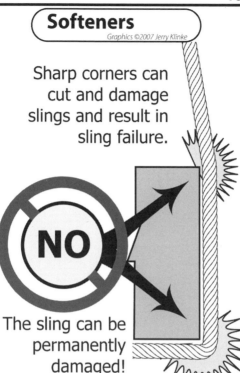

NO

The sling can be permanently damaged!

Applies to all types of slings!

But synthetic slings especially need protection from load edges - even if the edge is not "razor" sharp.

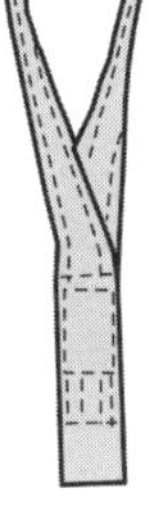

An edge without adequate protection may result in a catastrophic sling failure.

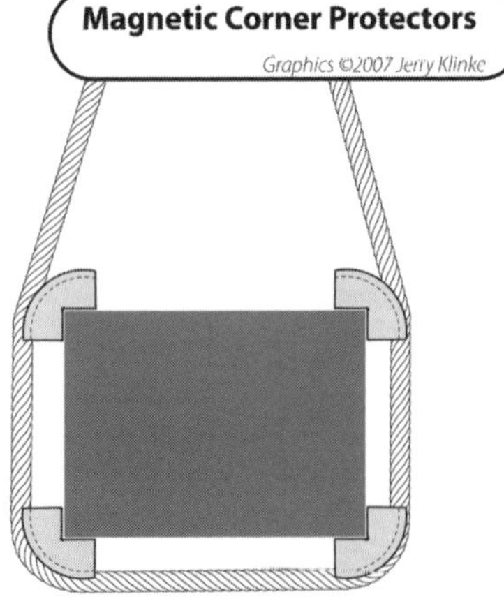
Magnetic Corner Protectors
Graphics ©2007 Jerry Klinke

These magnetic edge protectors are made out of nylon and can be used with both synthetic and wire rope slings.

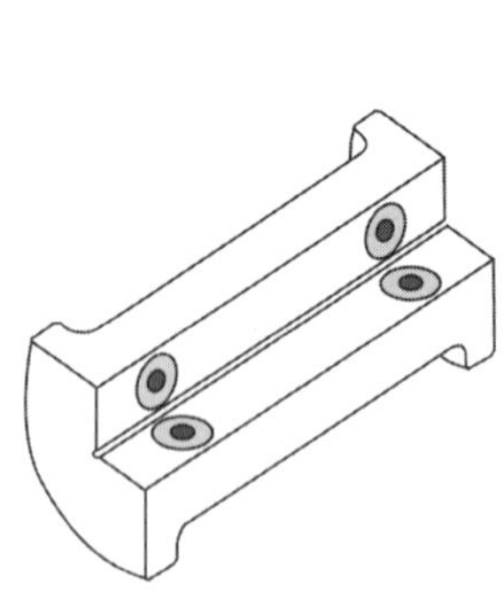

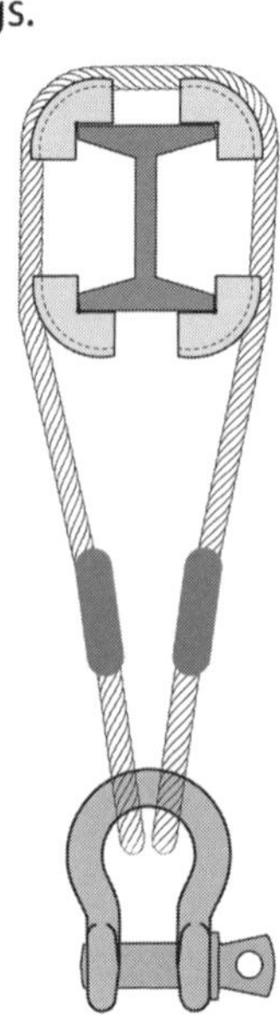

Available through Lift-It Manufacturing Company, Inc.
Phone: (323) 582-6076
website: www.lift-it.com

Misc Hazards

Graphics ©2007 Jerry Klinke

Wide WEB Sling Hazard

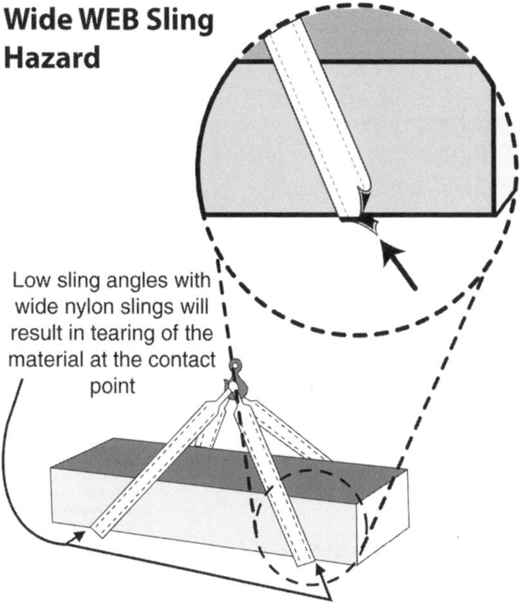

Low sling angles with wide nylon slings will result in tearing of the material at the contact point

Grading & Rigging

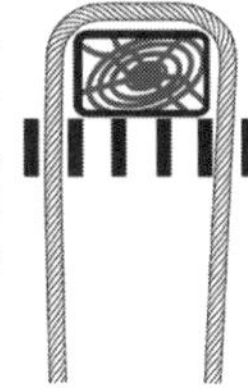

First, make sure the grading will structurally support the load (engineering analysis may be required) then use a piece of wood, metal pipes or rods to distribute the load and protect the sling from damage against the sharp edges of the grading.

Eyebolts

Graphics ©2007 Jerry Klinke

Eyebolts should be threaded into the surface a minimum of 1.5 times the thread diameter, this also applies to hoist rings.

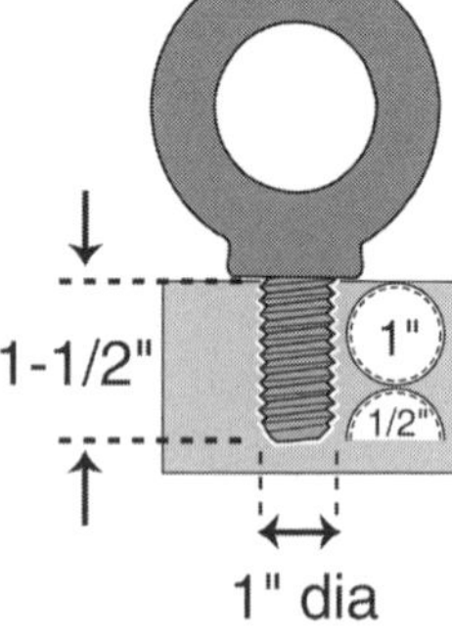
1-1/2"
1"
1/2"
1" dia

Insufficient thread engagement may cause the threads to fail

The condition of the threads is a key issue when using eyebolts. This applies to BOTH the eyebolt and the threads in the equipment being hoisted.

Always make sure the threads are clean and in good condition.

Inspect eye bolts before each use for defects that may effect proper thread engagement.

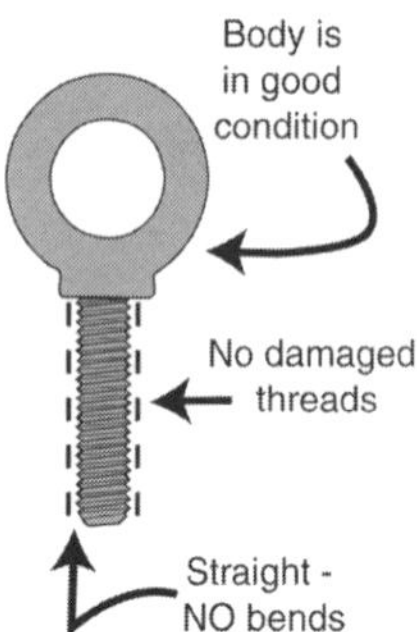
Body is in good condition
No damaged threads
Straight - NO bends

- Inspect and clean the eye bolt threads and the hole.
- Screw the eye bolt on all the way down and properly seat.
- Ensure the tapped hole in the body has a minimum depth of one-and-a-half times the bolt diameter.
- The shoulder should be in full contact with the surface of the object being lifted.

Eyebolts

Graphics ©2007 Jerry Klinke

YES *Only shoulder eyebolts can be used for angular lifting*. The shoulder must be flush and securely tightened against the load.*

* The working load limit (WLL) must be reduced in accordance with manufacturer's recommendations when shoulder eyebolts are used for angular lifting

YES *Eyebolts not shouldered shall only be used for in-line loads*

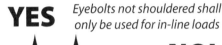

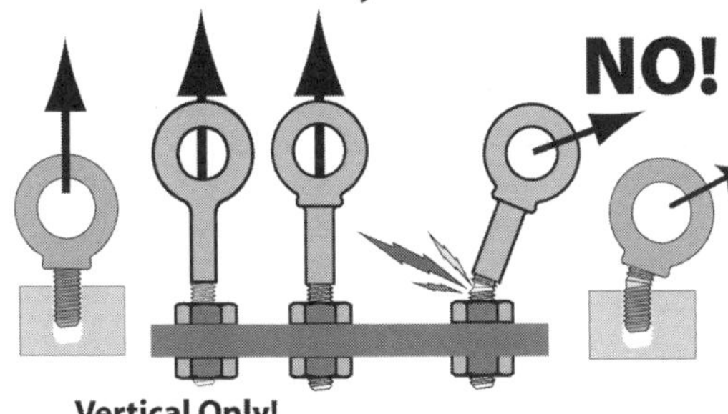
NO!
Vertical Only!

Be alert for situations that may leave the threaded shaft without side support and that would allow the shaft to bend. Some examples are when a counter-bored hole or multiple washers are used under the shoulder of the eyebolt.

NO!

Eyebolts

Graphics ©2007 Jerry Klinke

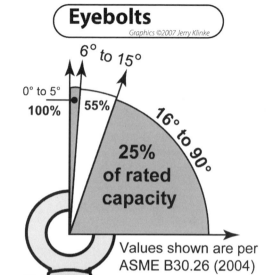

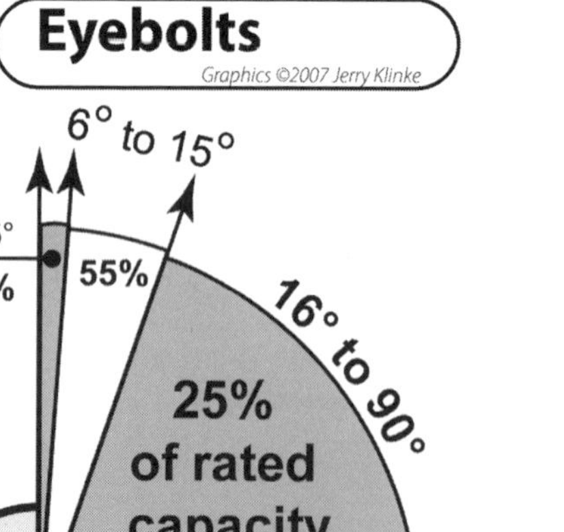
6° to 15°
0° to 5°
100%
55%
16° to 90°
25%
of rated
capacity
Values shown are per
ASME B30.26 (2004)

Shouldered eye bolts should only be used for angular lifting if rated by the eye bolt manufacturer.

NO!
YES

When using shoulder eyebolts for angular lifts, the plane of the eye MUST be aligned in the direction of loading.

Flat washers may be used under the shoulder to adjust the position of the plane of the eye (see page 81)

"Reeving"

Graphics ©2007 Jerry Klinke

WRONG

This illustration shows an extremely POOR rigging practice. By running or "reeving" a single sling through the eye bolts will immensely amplify the force on the eye bolts due to the resulting force.

Eyebolts will bend or break because of this tremendous force applied to them, this practice should ALWAYS be avoided!

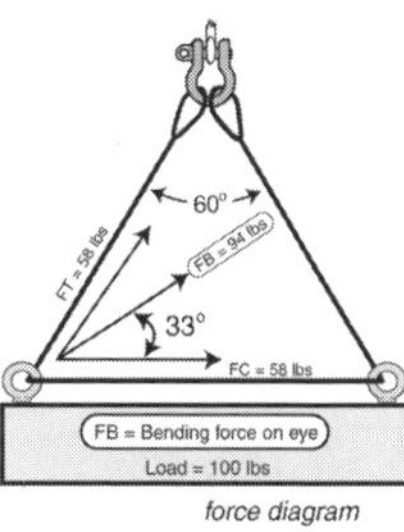
60°
FT = 58 lbs
FB = 94 lbs
33°
FC = 58 lbs
FB = Bending force on eye
Load = 100 lbs

force diagram

Shackle Use

Graphics ©2007 Jerry Klinke

WRONG

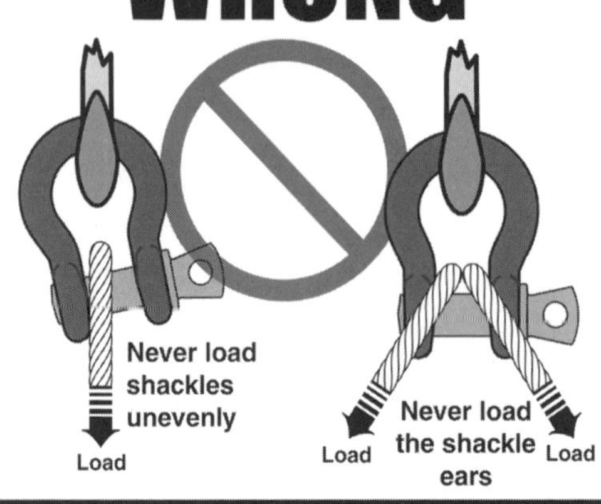

Never load shackles unevenly
Load
Never load the shackle ears
Load
Load

RIGHT

Load

Load

Washers keep the pin centered

120°

MAXIMUM Included Angle

Load

Load

Only the bow of a shackle is designed for angular loading

Shackle Use

Graphics ©2007 Jerry Klinke

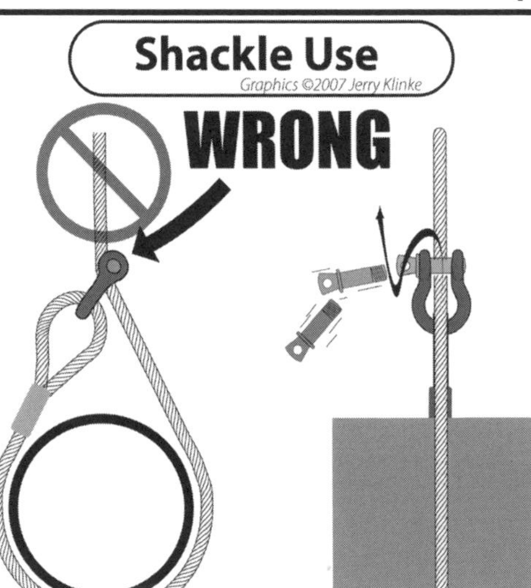
WRONG

"When a shackle is used in a choker hitch, the pin shall be connected to the eye of the sling"

ASME B30.26

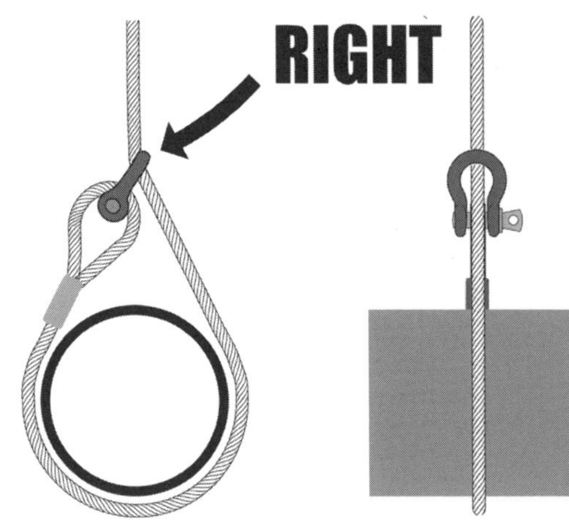
RIGHT

Shackle Use

WRONG

3/4"
Too small of a shackle size
1"

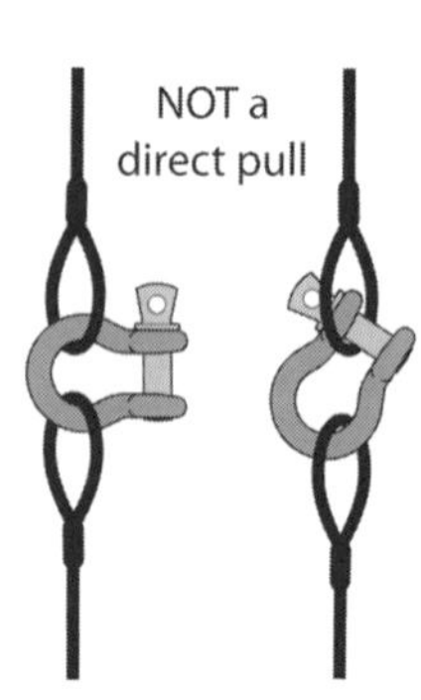
NOT a direct pull

RIGHT

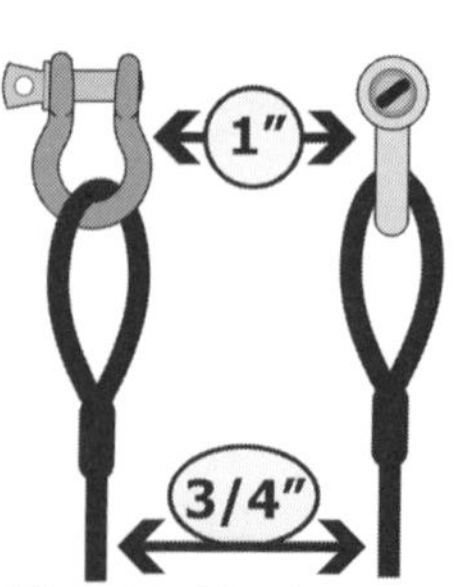
1"
3/4"
The shackle size must be larger than the wire rope size

Designed and rated for in-line loading

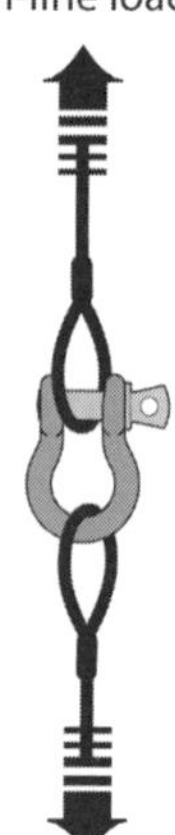

Shackle Use

Graphics ©2007 Jerry Klinke

How tight should you tighten a shackle pin ?

"The screw pin shall be fully engaged, with the shoulder (of the pin) in contact with the shackle body"

ASME B30.26

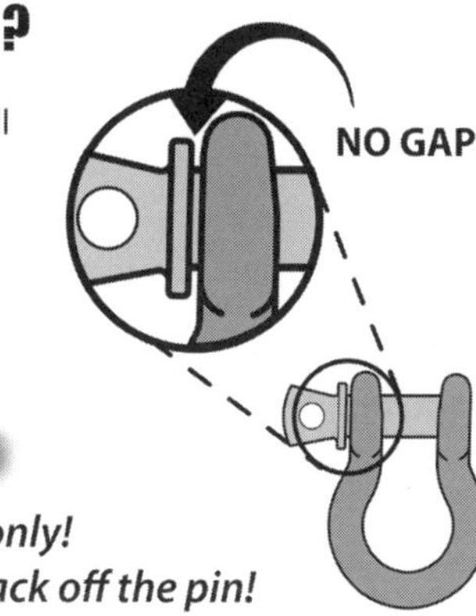
NO GAP!

Hand tighten only!

Don't back off the pin!

Point Loading of shackles:

"Point loading of Crosby® shackle bows and pins is acceptable as long as the load is reasonably centered"

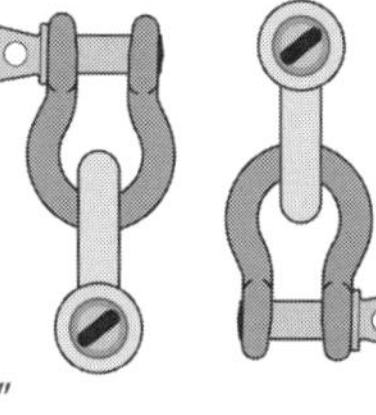

Note: Point loading is not specifically covered in the ASME standards. Contact other shackle manufacturers before point loading their shackles.

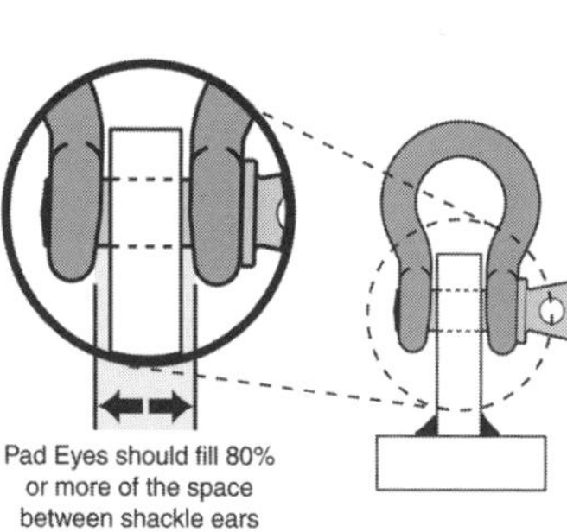

Pad Eyes should fill 80% or more of the space between shackle ears

Shackle Use

WEB slings and shackles

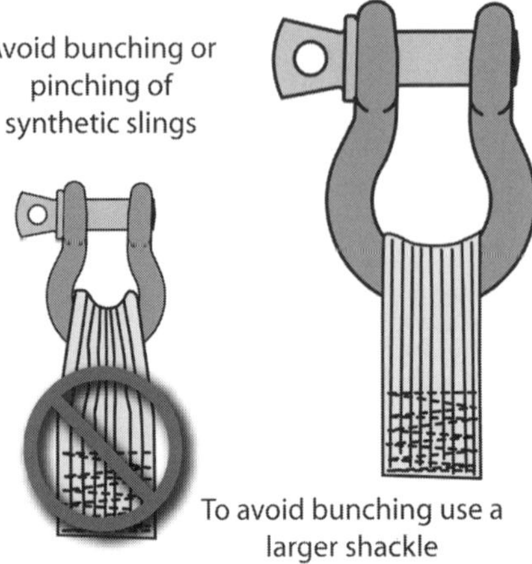
Avoid bunching or
pinching of
synthetic slings
To avoid bunching use a
larger shackle

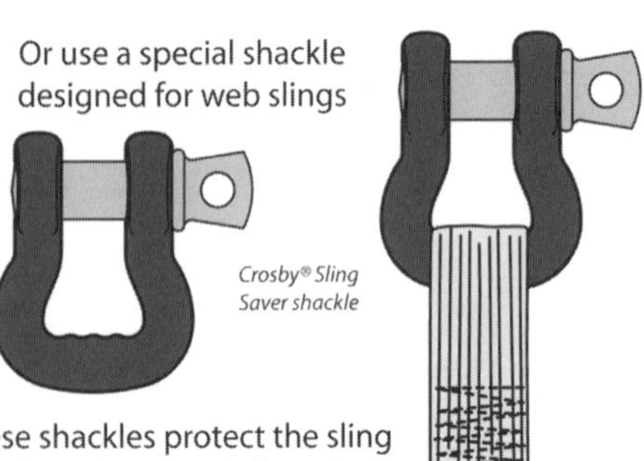
Or use a special shackle
designed for web slings
Crosby® Sling
Saver shackle

These shackles protect the sling from bunching, and allow the slings to be used at full capacity.

Shackle Use

Graphics ©2007 Jerry Klinke

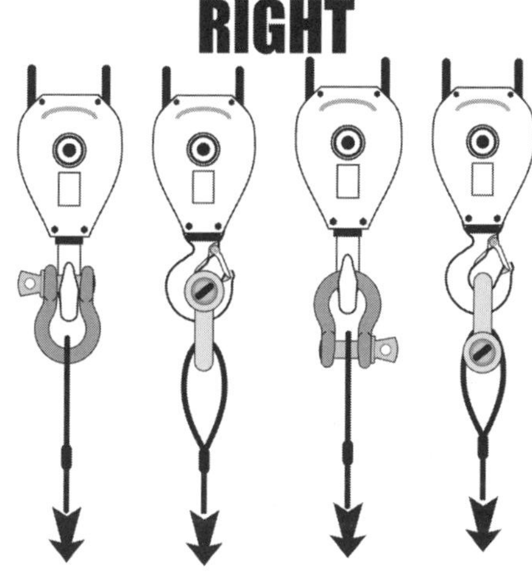
RIGHT

When a SINGLE leg sling is placed into a shackle, the shackle can be oriented with the pin up or down.

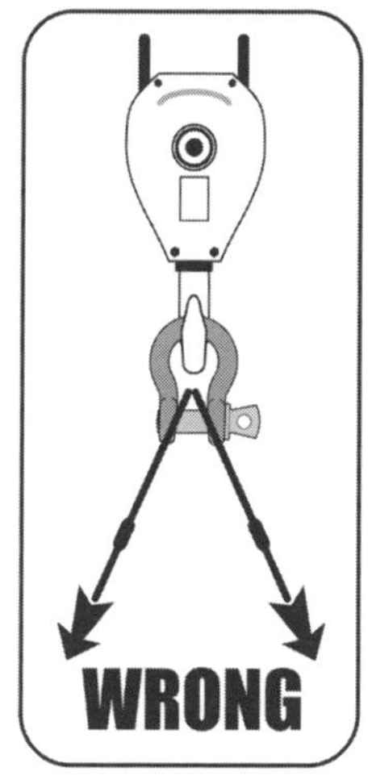
WRONG

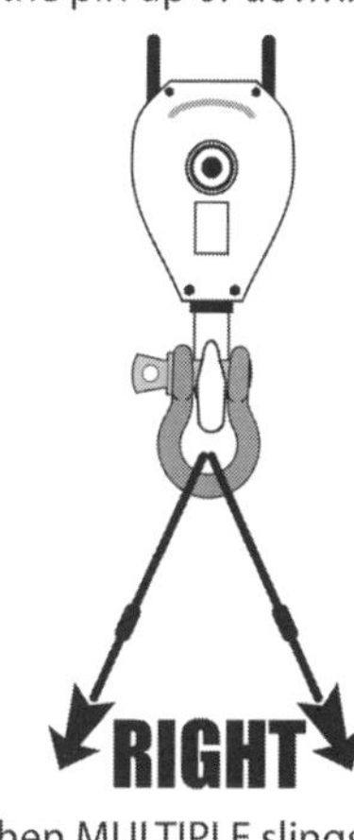
RIGHT

When MULTIPLE slings are placed into a shackle, the pin must always be UP.

Hooks

Graphics ©2007 Jerry Klinke

ASME B30.10 divides hooks into 2 groups:

Direct-pull

Hooks supporting a load in a direct-pull configuration, with the load carried in the base of the hook

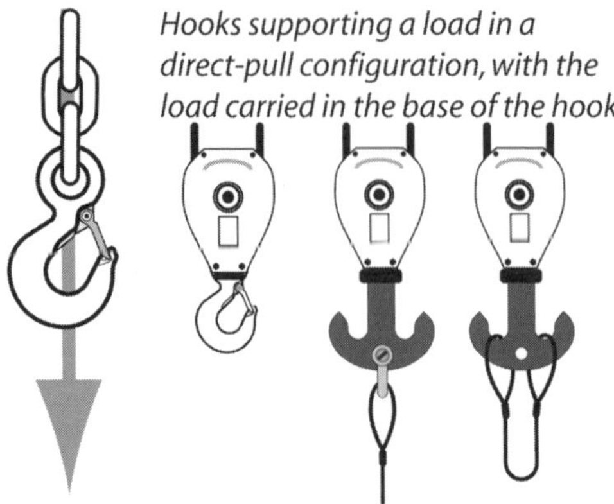

NOT in direct-pull

Hooks that do not support a load in a direct-pull configuration

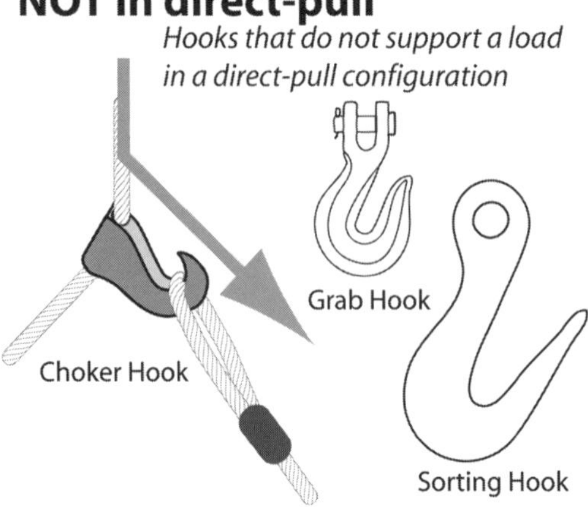
Grab Hook
Choker Hook
Sorting Hook

Not a comprehensive listing, refer to ASME B30.9 for specifics.

Hooks

Graphics ©2007 Jerry Klinke

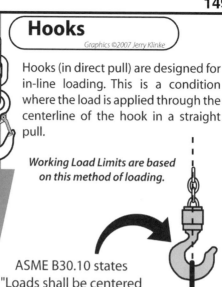

Hooks (in direct pull) are designed for in-line loading. This is a condition where the load is applied through the centerline of the hook in a straight pull.

Working Load Limits are based on this method of loading.

ASME B30.10 states "Loads shall be centered in the base of the hook to prevent point loading"

100%
of rating

The illustration below shows how quickly the capacity of the hook decreases with point loading.

88%
of rating
79%
of rating
71%
of rating
41%
of rating

DO NOT POINT LOAD HOOKS!

Hooks & Angles

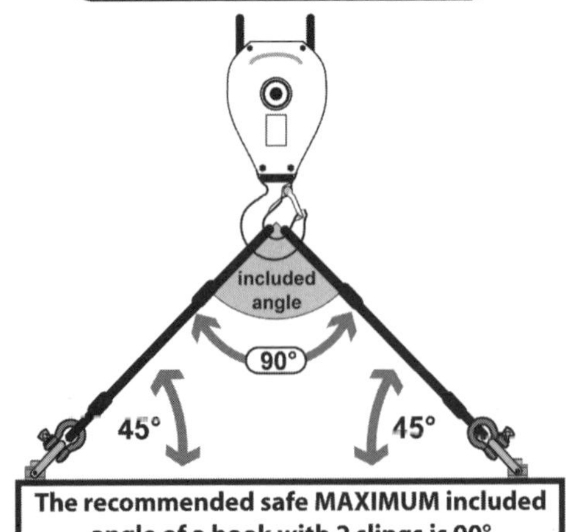
included
angle
90°
45°
45°

The recommended safe MAXIMUM included angle of a hook with 2 slings is 90°

Use a shackle when lower horizontal angles are encountered.

Shackles & Angles

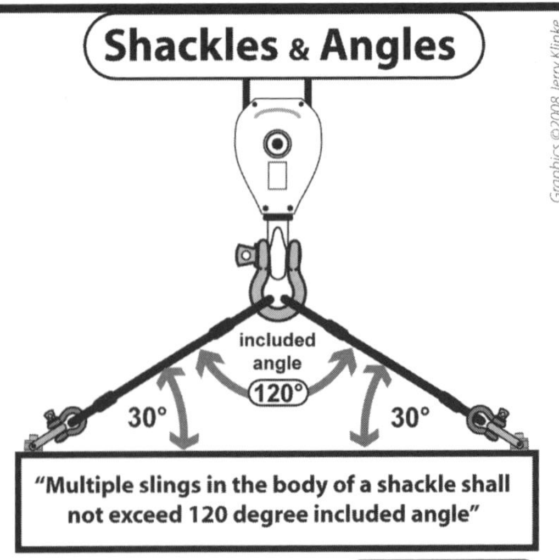
included
angle
120°
30°
30°

"Multiple slings in the body of a shackle shall not exceed 120 degree included angle"

ASME B30.9

Hooks

Graphics ©2007 Jerry Klinke

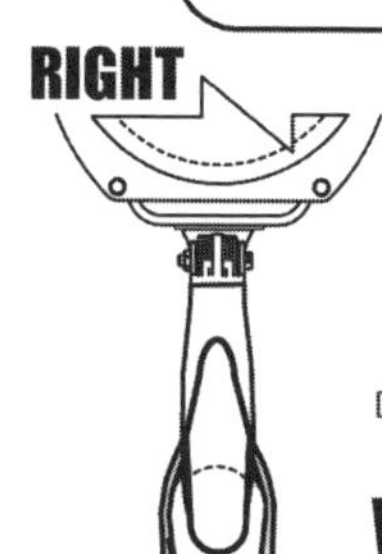
RIGHT

"an object in the eye of a (synthetic) sling should not be wider than one-half the length of the eye"

ASME B30.9

Don't force the eye on the hook - get a larger sling!

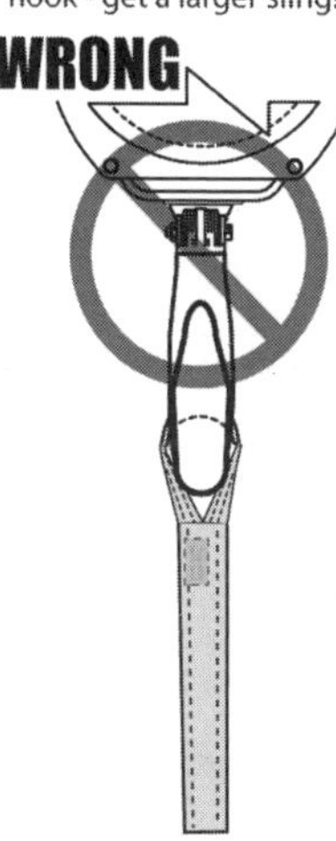
WRONG

"the width of the hook should not be wider than one half the length of the eye"

ASME B30.9

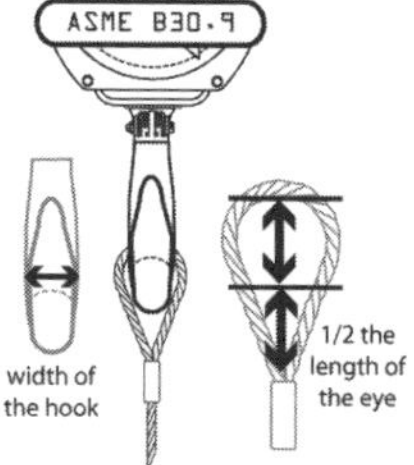
width of
the hook
1/2 the
length of
the eye

Duplex hooks should be loaded equally on both sides

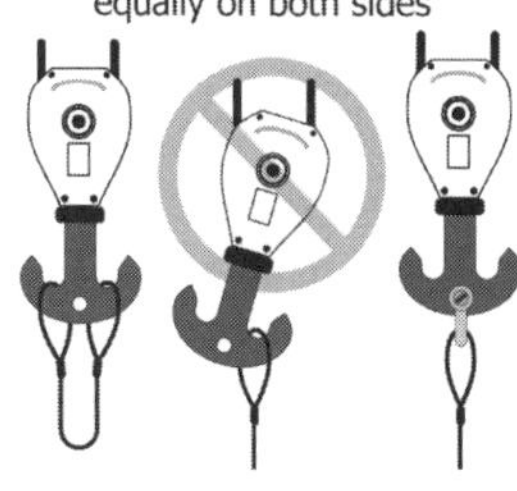

Hooks

Graphics ©2007 Jerry Klinke

Back loading
Hook
NO !

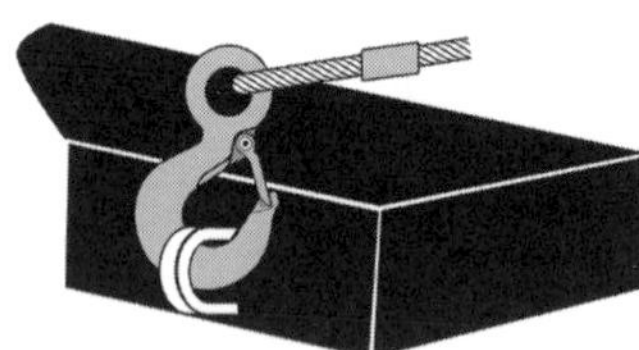

Side loading
Hook

NO !

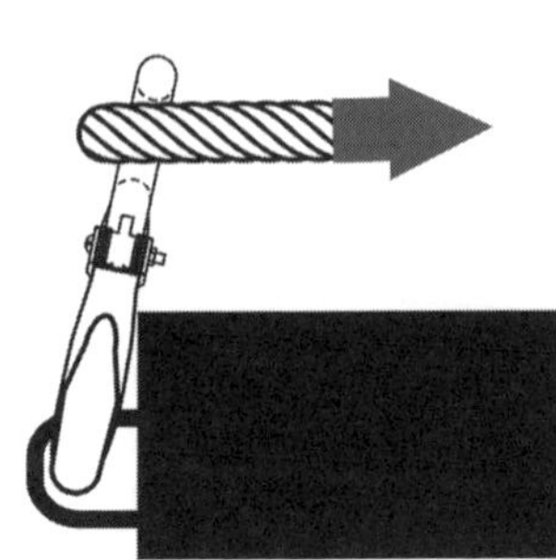

Hook Latches

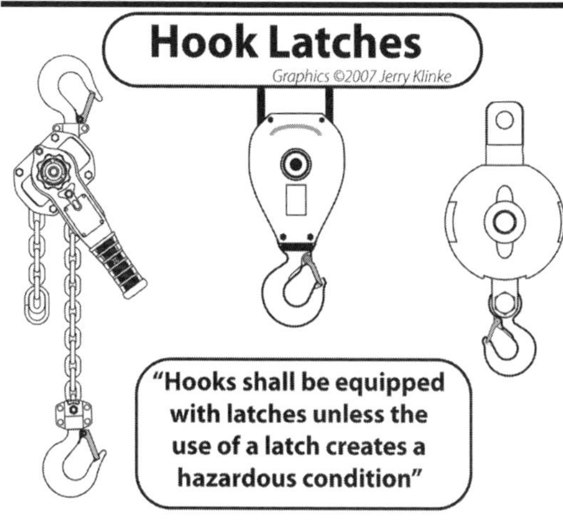
Graphics ©2007 Jerry Klinke
"Hooks shall be equipped with latches unless the use of a latch creates a hazardous condition"

If the hook has a latch – it MUST function properly!

ASME B30.9 does not specifically require latches on slings with hooks. It DOES require latches, if present, must function and seat properly.

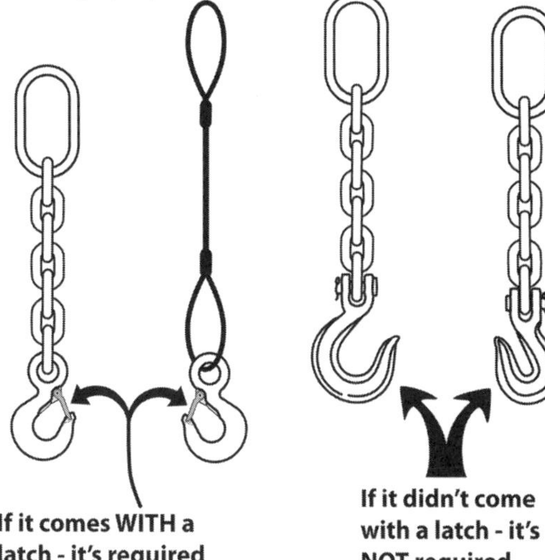
If it comes WITH a latch - it's required
If it didn't come with a latch - it's NOT required

Connecting Slings

Graphics ©2007 Jerry Klinke

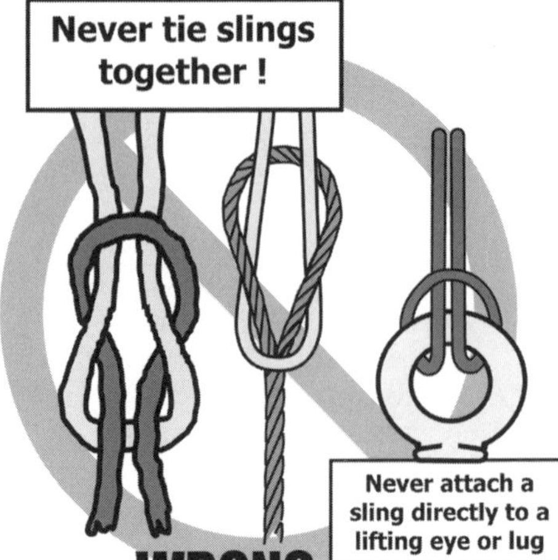
Never tie slings together !
Never attach a sling directly to a lifting eye or lug
WRONG

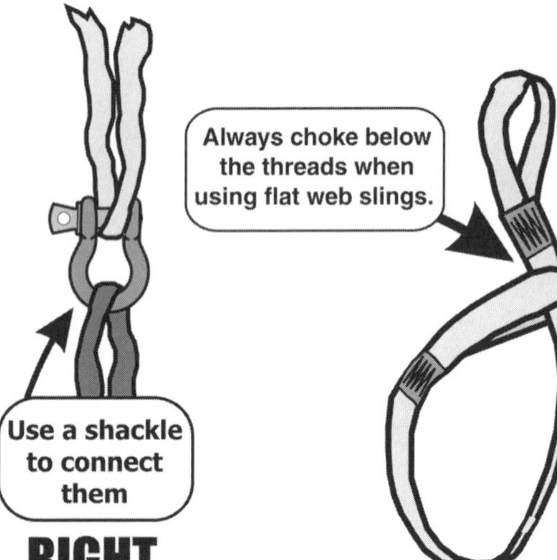
Always choke below the threads when using flat web slings.
Use a shackle to connect them
RIGHT

Rigging Concerns

Graphics ©2007 Jerry Klinke

NO !

C-Clamps should NEVER be used for overhead lifting applications!

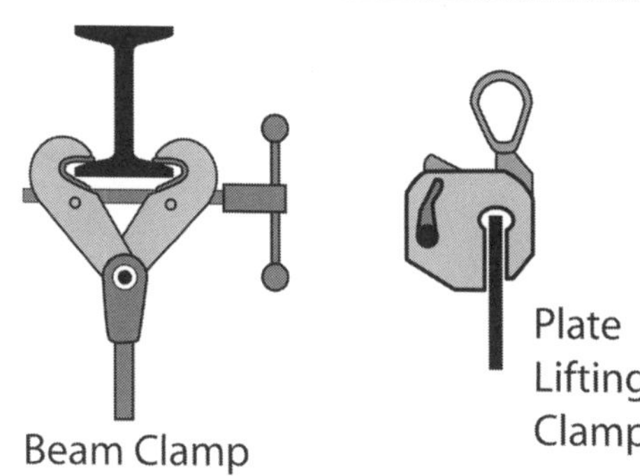
Beam Clamp
Plate Lifting Clamp

Use the proper lifting devices that are rated and approved for this type of lifting.

Hand Hoists

Graphics ©2007 Jerry Klinke

LEVER HOISTS (Come-a-longs)

- This equipment should only be operated and maintained by a competent person
- Do not exceed the rated capacity of the lever hoist
- Do not use the load chain as a sling
- Do not extend the operating lever
- Do not use undue effort to operate the lever hoist
- Do not use for lifting people
- Do not place people or body parts under the load

WRONG !

Come-a-longs are designed so that ONE average sized person can use it WITHOUT any cheaters or pipes. Most come-a-longs are designed and rated this way, so if any additional leverage is used - the hoist will be used beyond its safe rated capacity.

DON'T
DO THIS!
Use a larger capacity
hoist instead!

Hand Hoists

Graphics ©2007 Jerry Klinke

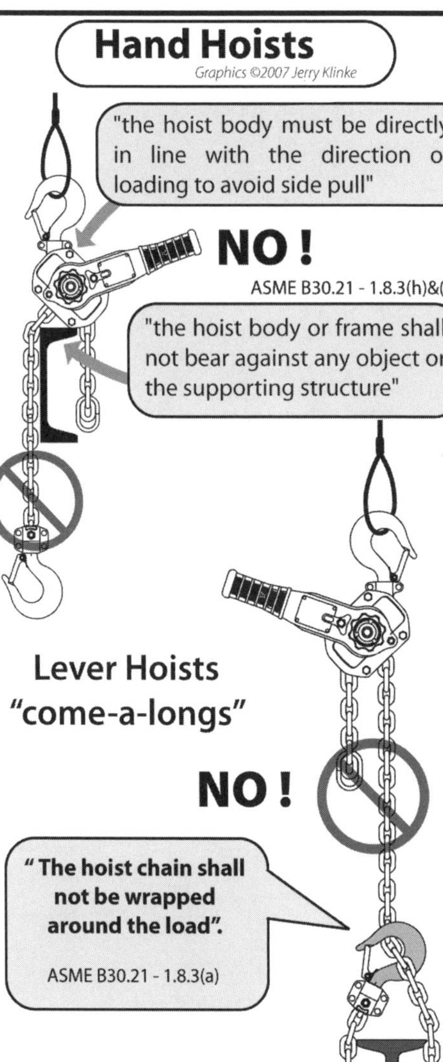
"the hoist body must be directly in line with the direction of loading to avoid side pull"
NO !
ASME B30.21 - 1.8.3(h)&(i)
"the hoist body or frame shall not bear against any object or the supporting structure"
Lever Hoists
"come-a-longs"
NO !
" The hoist chain shall not be wrapped around the load".
ASME B30.21 - 1.8.3(a)

Rigging Concerns

Tall loads not properly supported can shift suddenly.

Be aware of soft, unstable ground surfaces.

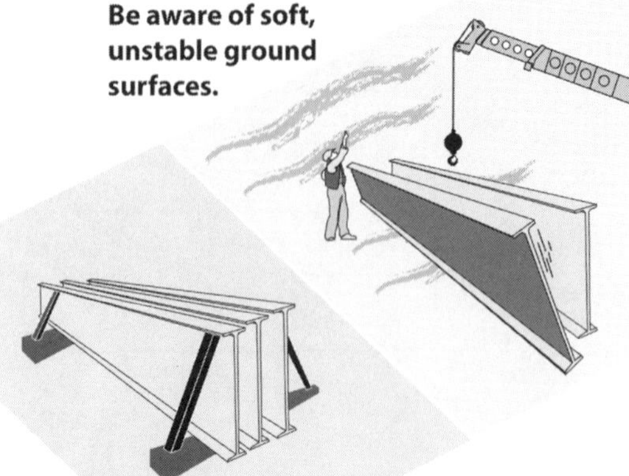

Always use proper cribbing and bracing to prevent load shifting.

If the load has a potential for binding during the lifting process, the use of a load cell to measure the force applied will help identify when the load has become stuck. This will prevent the overloading of the rigging and lifting equipment.

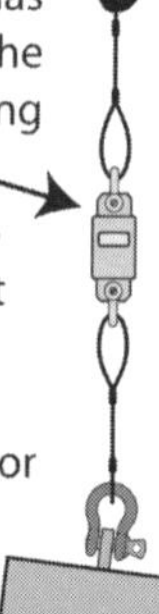

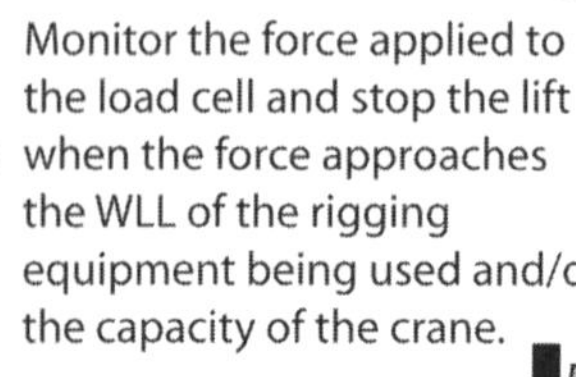

Monitor the force applied to the load cell and stop the lift when the force approaches the WLL of the rigging equipment being used and/or the capacity of the crane.

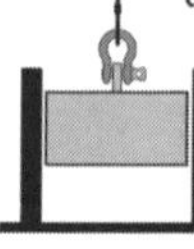

Center of Gravity

Graphics ©2007 Jerry Klinke

UNSAFE!

CORRECT
Center of Gravity is below the lift points

UNSTABLE!
Center of Gravity is above the lift points

The center of gravity is one of the most important concepts the rigger must understand.

The center of gravity is the point at which all of the weight of a load can be considered to be concentrated. It always acts vertically downward to bring the load to a position of equilibrium.

As long as the center of gravity is below the lift points, the object being lifted will be stable.

*But when the center of gravity is placed **above** the lift points, gravity will take over and shift the load in an effort to move the center of gravity below the lift points.*

THIS MEANS THE LOAD WILL MOST LIKELY BE DROPPED.

Center of Gravity

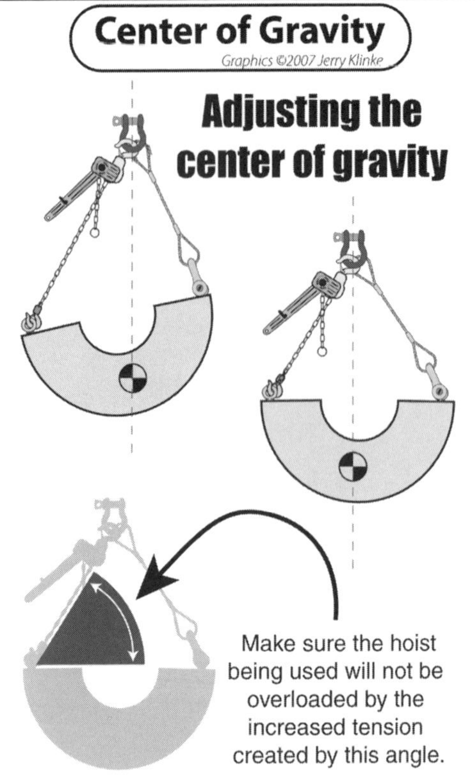
Graphics ©2007 Jerry Klinke
Adjusting the center of gravity
Make sure the hoist being used will not be overloaded by the increased tension created by this angle.

In order to make a level lift, the crane hook must be directly above the center of gravity, and the slings being used are of the proper length AND attached on or above the center of gravity. Hoists or come-a-longs are sometimes used to adjust the CG, but caution must be exercised to avoid exceeding the WLL of the hoist. Refer to the section in this book on calculating tension on slings.

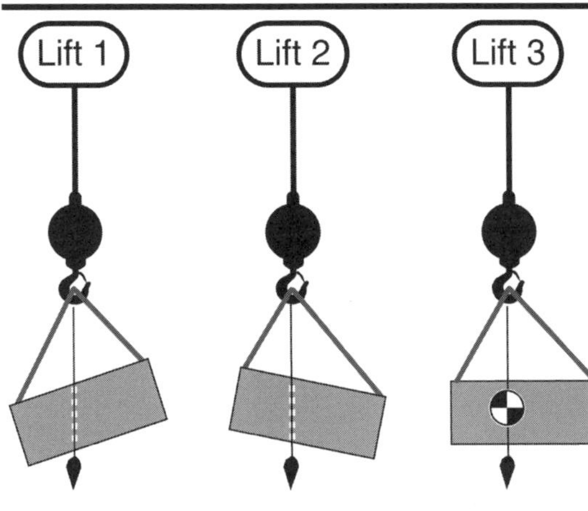
Lift 1
Lift 2
Lift 3

Using the trial and error method for finding the center of gravity.

This is the LEAST desirable method to use; extreme caution must be taken to ensure that the slings and hardware are not overloaded.

Lift 1: Raise the load and use a plumb line to mark a vertical line directly below the hook.

Lift 2: Change the sling lengths and raise and mark another vertical line directly below the hook.

The intersection of the two lines will indicate the center of gravity. Move the hook directly over the center of gravity, then attach the proper length slings.

In this book we only present a few basic concepts on the center of gravity. Difficult lifts should be analyzed by a competent engineering professional prior to lifting.

SLINGS HARDWARE FORMULAS EXAMPLES REFERENCE STANDARDS EXTRAS

Some key points to follow are:

1. Determine both the location and the effect of the center of gravity.
2. Always attach the slings ABOVE the center of gravity.
3. Try to visualize the entire lift and turning operation.
4. If you are not confident about the procedure, consult with a competent engineer

Referring to the illustration on the right, at the start of the lift make sure the attachment is as high above the center of gravity as possible. Try to keep the sling tight against the load.

Step 2 requires the hoist to be raised SLOWLY and moved slightly towards the known center of gravity. When the block is tilted, the center of gravity is moved between the point of contact and the point of attachment.

To avoid having the load "fall" the attachment points must be carefully positioned. The hitch must always be as high as possible above the center of gravity. The attachment point should be located as close to the vertical line of the center of gravity as possible to minimize the turning moment. One edge of the load acts as a pivot point, around which the turn is made.

When the load is turned, the crane is moved to prevent sliding the pivot edge along the floor.

Single Hook Turning

Graphics ©2007 Jerry Klinke

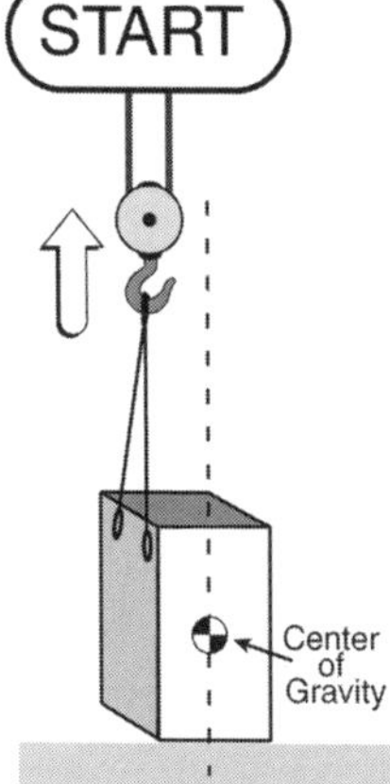
START
Center of Gravity

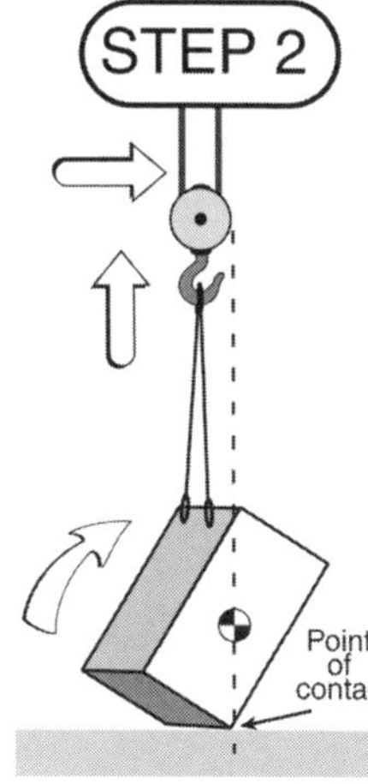
STEP 2
Point of contact

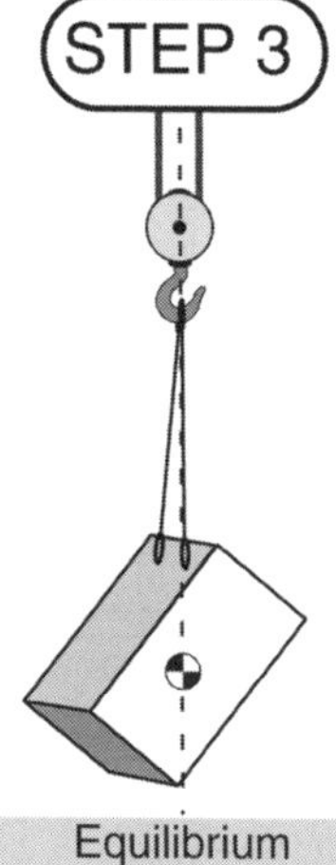
STEP 3
Equilibrium

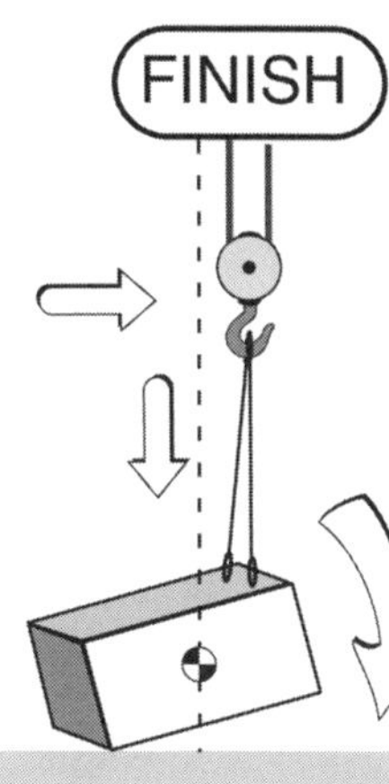
FINISH

Turning is often necessary to position a component to obtain the best setup for working, or to position it for assembly. Unfortunately, more trouble is encountered when turning with a single hook because few loads offer ideal attachment points, or have uniform center of gravity as shown in the example above.

Two Hoist Turning

Graphics ©2007 Jerry Klinke

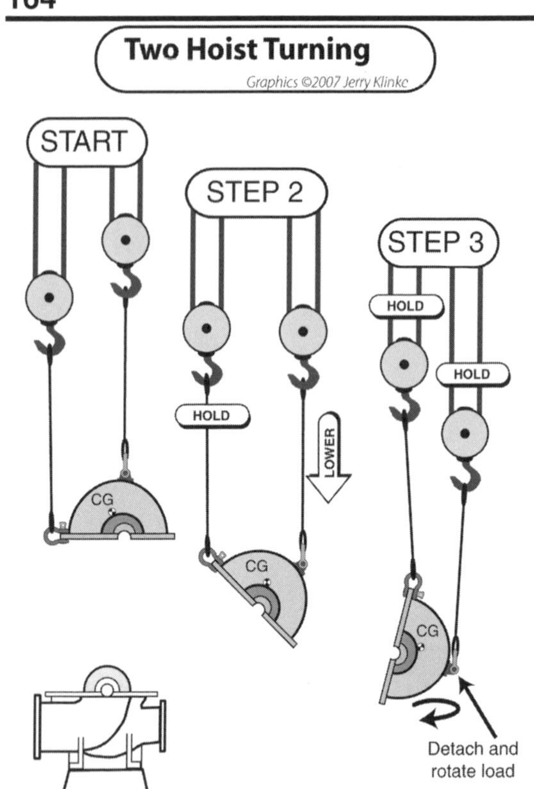
START
STEP 2
STEP 3
CG
HOLD
LOWER
CG
HOLD
HOLD
CG
Detach and rotate load

Two Hoist turning is used for turning loads freely in the air while supported. This is the most difficult type of rigging operation and should be done only after carefully planning the operation. You must also be aware that one hoist must have sufficient capacity to lift the entire load, because at one point in the move the load must be freely suspended from only one hoist.

Two Hoist Turning

Graphics ©2007 Jerry Klinke

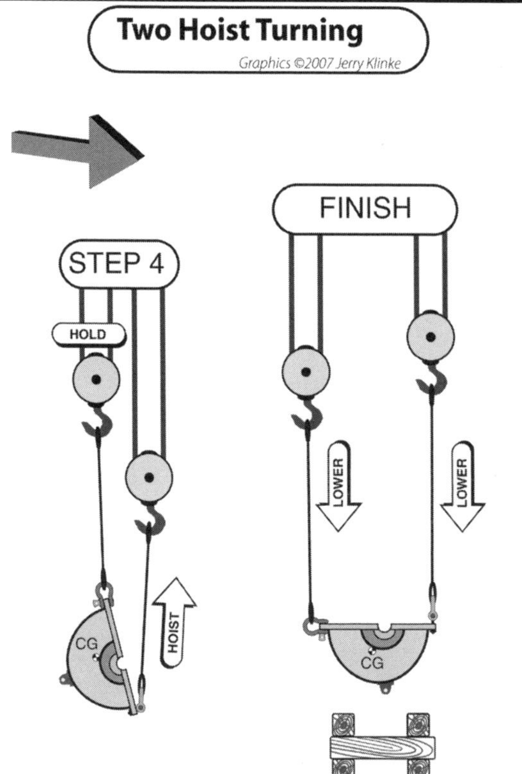

FINISH
STEP 4
HOLD
LOWER
LOWER
HOIST
CG
CG

One sling on the main hoist supports the load and acts like a pivot around which the turn is made. A second sling on the auxiliary hoist is used to provide load control.

Note: It will be necessary to disconnect the auxiliary hoist prior to turning in the air, and then to reconnect it after the turn has been made. Also be aware, and plan for any changes that may be required to the attachment points during this operation. The hardware (eye bolt, shackle, etc.) on the main hoist must be rated to handle the drastic changes in tension that will occur

Turning w/choker

Graphics ©2007 Jerry Klinke

When turning with a choker hitch, it is extremely important to remember that the eye of the cable must be placed so it faces in the opposite direction of the turn. If this is not done, the load will be turned in a loose cable.

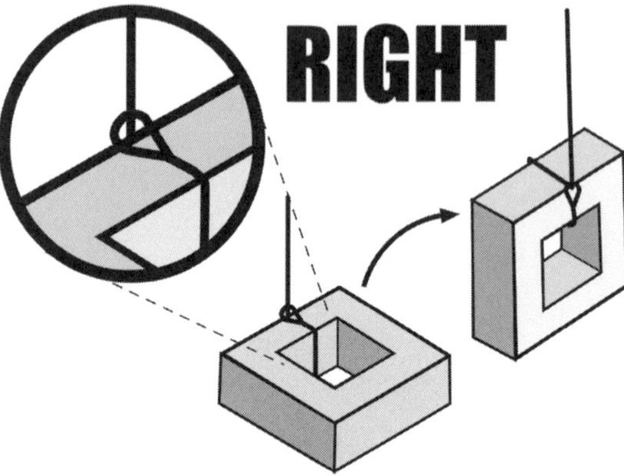
RIGHT

Remember that the capacity of the sling will be reduced because of the "tight" choke angle formed when the load is turned.

WRONG

Shock Loading

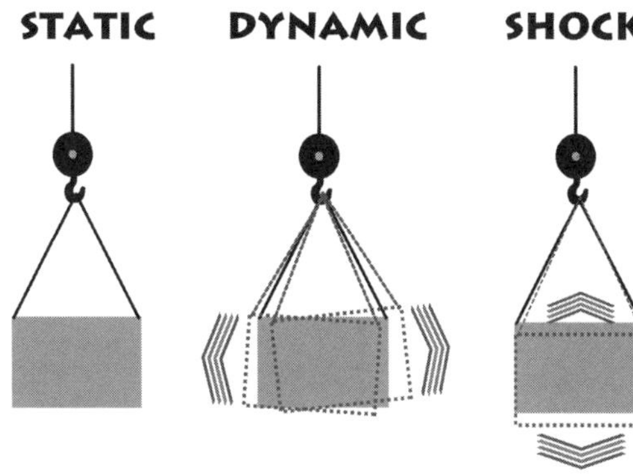
STATIC
DYNAMIC
SHOCK

When the load is suspended and not moving it is called a static load. When the load begins to move, additional stresses are imposed on the rigging, this is called a dynamic load. Shock Loads (considered a Dynamic load) act suddenly upon the rigging. Their effects can be quite disastrous because of the tremendous additional tension to the rigging. Unless the rigging was considerably larger than what was needed for the static load, failure of the rigging is possible.

When lifting a load, the crane hook should be started very slowly until the sling becomes taut. Then continue lifting slowly until the load is suspended. The speed with which you lift or lower the crane hook should increase or decrease gradually. Any sudden starts or stops place a much heavier load on the slings. How much the load is increased on the slings is impossible to tell. It could be from 2 times the actual load weight, or 100 times the actual load weight.

!WARNING!

Once any sling has been shock loaded it MUST be removed from service, re-inspected and re-tested to determine if it has been damaged.

Knots

Graphics ©2007 Jerry Klinke

BOWLINE

A bowline is a very secure knot which won't slip, regardless of the load applied.

IMPORTANT: These knots are for rope only. NEVER tie ANY knots in rigging used for industrial rigging and overhead lifting applications.

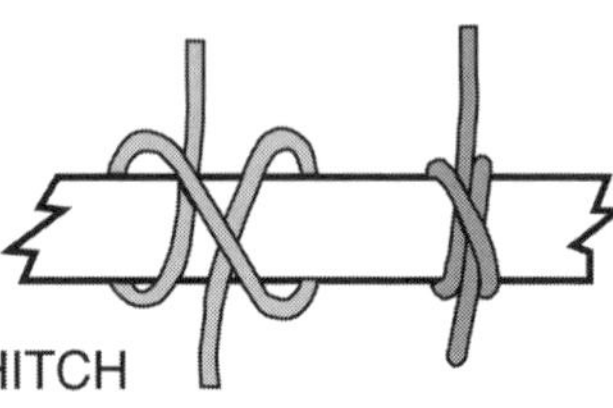

CLOVE HITCH

The clove is a popular knot for securing to posts, bars, and other round objects.

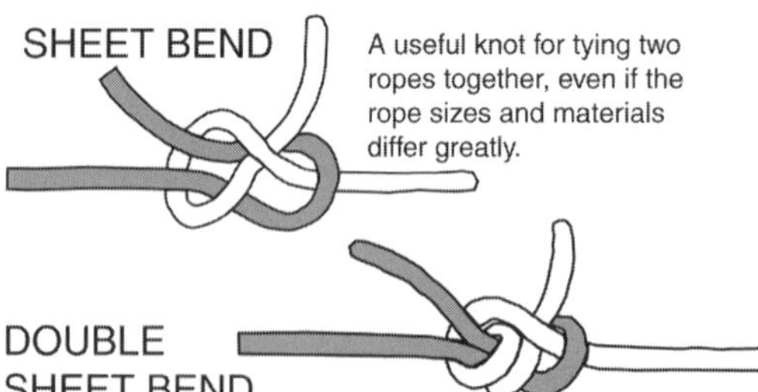

SHEET BEND

A useful knot for tying two ropes together, even if the rope sizes and materials differ greatly.

DOUBLE SHEET BEND

This knot provides greater security, especially in plastic rope. Its the same as the sheet bend but with an extra coil around the standing loop.

Many years ago, before the advent of high strength synthetic materials, ropes were commonly used to rig and lift heavy equipment. Knots became a very important part of rigging knowledge because an improperly tied knot would reduce the capacity of the rope by as much as 90%.

Knots

Graphics ©2007 Jerry Klinke

Square Knot

Granny Knot

Bowline on a bight

Sheepshank Knot

Two Half Hitches

Today in construction and industry rope is mainly used for tag lines, or lifting lighter items by hand. However, rope is still used today by rescue crews and mountain climbers. The examples shown are just a few of the hundreds of knots that have seen generations of use.

Double Wrap Hitches

Graphics ©2007 Jerry Klinke

To improve a choker or basket hitch, consider using a double wrap. This has better control of the load and does not change the rated capacity of the hitch. Make sure that the wraps do not overlap at the bottom of the hitch when using this type of hitch. Be aware that it may also crush the load if the load does not have structural integrity.

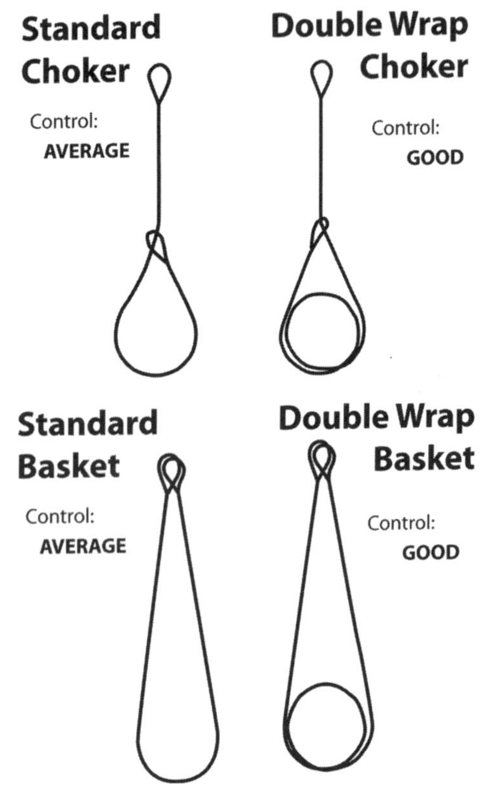
Standard Choker
Control: AVERAGE
Double Wrap Choker
Control: GOOD
Standard Basket
Control: AVERAGE
Double Wrap Basket
Control: GOOD

REFERENCE

Reference

To find an unknown side of a right angle triangle:

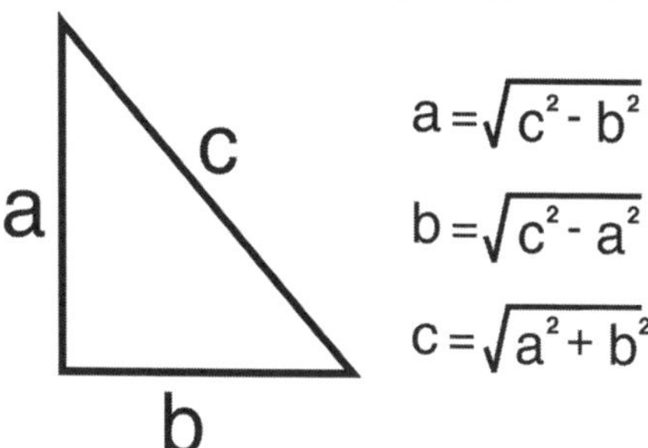
a
b
c
$a = \sqrt{c^2 - b^2}$
$b = \sqrt{c^2 - a^2}$
$c = \sqrt{a^2 + b^2}$

UNITS OF MEASURE

1 US ton (short)	=	2000 lbs
1 US ton (short)	=	.91 metric ton
1 US ton (short)	=	907 kgs
1 metric ton	=	2204.62 lbs
1 metric ton	=	1.102 US tons
1 metric ton	=	1000 kgs
1 US pound (lb)	=	16 ounces
1 US pound (lb)	=	.45 kg
1 kilogram (kg)	=	2.2 lbs
1 kilogram (kg)	=	35 ounces
1 kilogram (kg)	=	1000 grams
1 US (liq) gallon	=	4 quarts
1 US (liq) gallon	=	3.8 liters
1 liter	=	.264 gallons (US)
1 liter	=	1.06 quarts
1 US gallon water	=	8.3 lbs
1 cubic ft of liquid	=	7.5 US gallons

Schedule 40
pipe weight chart

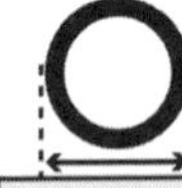

Nominal Size	OD	ID	Weight per foot
1/2"	0.840	0.622	0.851
3/4"	1.050	0.824	1.131
1"	1.315	1.049	1.679
1-1/4"	1.660	1.380	2.273
1-1/2"	1.900	1.610	2.718
2"	2.375	2.067	3.653
2-1/2"	2.875	2.469	5.793
3"	3.500	3.068	7.576
3-1/2"	4.000	3.548	9.109
4"	4.500	4.026	10.79
4-1/2"	5.000	4.506	12.53
5"	5.563	5.047	14.62
6"	6.625	6.065	18.970
8"	8.625	7.981	28.550
10"	10.750	10.020	40.480
12"	12.750	11.938	49.560
14"	14.000	13.124	63.440
16"	16.000	15.000	82.770

Schedule 80

pipe weight chart

Nominal Size	OD	ID	Weight per foot
1/2"	0.840	0.546	1.09
3/4"	1.050	0.742	1.47
1"	1.315	0.957	2.17
1-1/4"	1.660	1.278	3.00
1-1/2"	1.900	1.500	3.63
2"	2.375	1.939	5.02
2-1/2"	2.875	2.323	7.66
3"	3.500	2.900	10.25
3-1/2"	4.000	3.364	12.50
4"	4.500	3.826	14.98
4-1/2"	5.000	4.290	17.61
5"	5.563	4.813	20.78
6"	6.625	5.761	28.57
8"	8.625	7.625	43.39
10"	10.750	9.750	64.43
12"	12.750	11.376	88.63
14"	14.000	12.500	106.13
16"	16.000	15.000	136.61

Solid round stock

Steel only

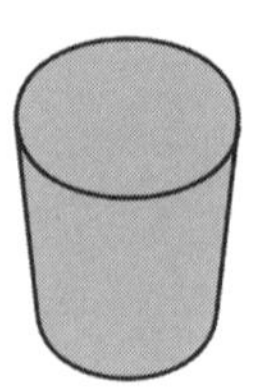

Diameter	Weight per foot
1/2"	0.668
3/4"	1.502
1"	2.670
1-1/4"	4.173
1-1/2"	6.008
2"	10.680
2-1/2"	16.690
3"	24.03
3-1/2"	32.710
4"	42.730
4-1/2"	54.080
5"	66.760
6"	96.130
7"	130.800
8"	170.900
9"	216.300
10"	267.000

Sheets / Plates

Steel only

Thickness	lbs. per sq. ft.
3/16"	7.66
1/4"	10.21
5/16"	12.76
3/8"	15.32
7/16"	17.87
1/2"	20.42
9/16"	22.97
5/8"	25.53
11/16"	28.08
3/4"	30.63
7/8"	35.74
1"	40.84
2"	81.68
3"	122.52
4"	163.36
5"	204.2
6"	245.04
7"	285.88
8"	326.72

Flat Strips

Steel only

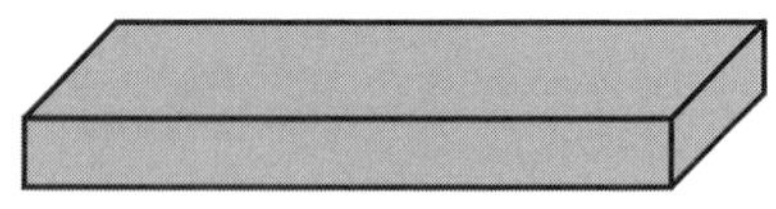

	Thickness			
	Weight per foot			
Width	1/4"	3/8"	1/2"	3/4"
1"	0.85	1.28	1.70	2.55
2"	1.70	2.55	3.40	5.10
3"	2.55	3.83	5.10	7.65
4"	3.40	5.10	6.80	10.20
5"	4.25	6.38	8.50	12.75
6"	5.10	7.65	10.20	15.30
7"	5.95	8.93	11.90	17.85
8"	6.80	10.20	13.60	20.40

Timber Beams

Capacity of Yellow Pine beams

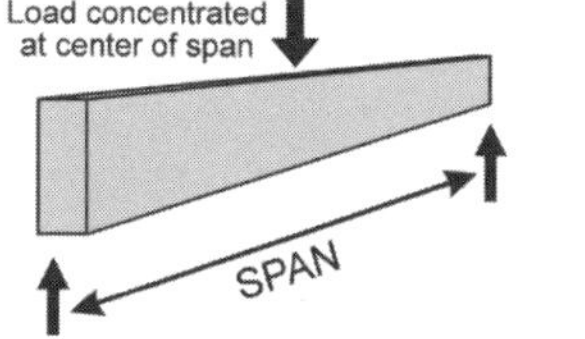
Load concentrated
at center of span
SPAN

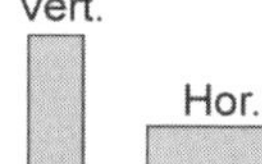
Vert.
Hor.

SIZE of Timber	Span, in feet						
	4	6	8	10	12	14	16
	Support load in lbs						
4 x 4	990	650	480	380	310	260	220
4 x 6 Hor.	1530	1010	750	590	480	400	340
6 x 6	3440	2290	1700	1340	1110	930	800
6 x 8 Hoz.	4700	3120	2320	1830	1510	1270	1090
6 x 8 Vert.	6430	4260	3180	2520	2080	1760	1520
8 x 8	6690	6660	4330	3440	2840	2400	2070
10 x 10	10740	10690	10640	7020	5800	4930	4270
12 x 12	15690	15620	15550	12490	10340	8790	7630
14 x 14	21630	21520	21420	21320	16780	14290	12410

Timber Column

Yellow Pine

These charts are based on yellow pine of fir timber in first class condition. Weight transfer is based on the post bearing surface distribuded over a greater area of yellow pine or other suitable material. Dimensions are for rough lumber, not surfaced.

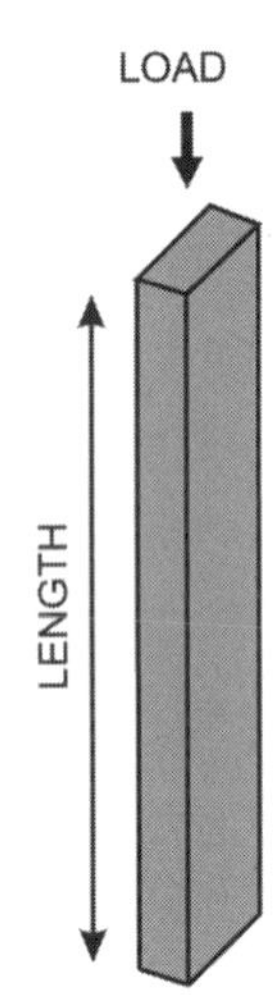
LOAD
LENGTH

SIZE	Length, in feet						
	8	10	12	14	16	18	20
	Load in TONS						
4 x 4	4.7	3.9					
6 x 6	13.4	12.2	10.9	9.8	8.6	7.5	6.3
8 x 8	27.3	25.6	23.9	22.4	20.8	19.2	17.5
10 x 10	45.9	43.9	41.9	39.8	37.7	35.7	33.6
12 x 12	69.4	66.9	64.5	61.9	59.4	56.9	54.5
14 x 14	97.7	94.8	91.8	88.9	85.9	83.1	80.2

Rebar
Weight Estimator

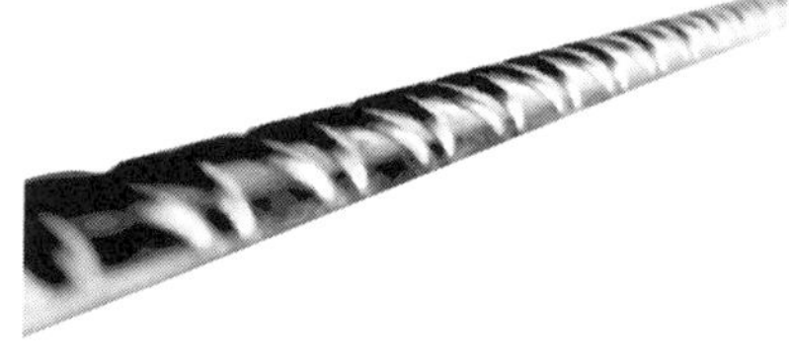

Bar No.	Diameter (inches)	Weight per foot
3	0.375	0.376
4	0.500	0.668
5	0.625	1.043
6	0.750	1.502
7	0.875	2.044
8	1.000	2.670
9	1.128	3.400
10	1.270	4.303
11	1.410	5.313
14	1.693	7.650
18	2.257	13.600

Steel S-type I-Beams

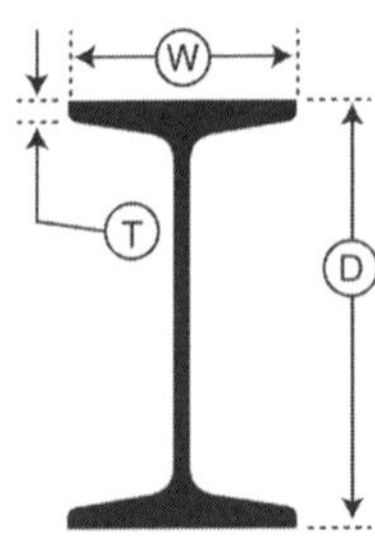
W
T
D

Section number	Wt. Per foot (lbs)	Dimensions		
		D	**W**	**T**
S4	7.7	4	2 5/8	5/16
	9.5	4	2 3/4	5/16
S5	10	5	3	5/16
S6	12.5	6	3 3/8	3/8
	17.25	6	3 5/8	3/8
S7	15.3	7	3 5/8	3/8
S8	18.4	8	4	7/16
	23	8	4 1/8	7/16
S10	25.4	10	4 5/8	1/2
	35	10	5	1/2
S12	31.8	12	5	9/16
	35	12	5 1/8	9/16
	40.8	12	5 1/4	11/16
	50	12	5 1/2	11/16

Data supplied for reference only. Dimensions and weights may be approximated, contact your steel supplier for specific details.

Steel S-type I-Beams *Continued*

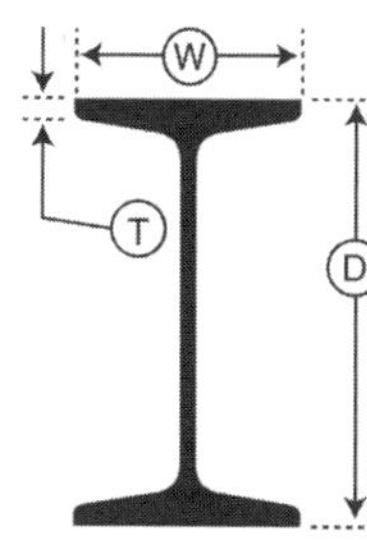
W
T
D

Section number	Wt. Per foot (lbs)	Dimensions		
		D	W	T
S15	42.9	15	5 1/2	5/8
	50	15	5 5/8	5/8
S18	54.7	18	6	11/16
	70	18	6 1/4	11/16
S20	66	20	6 1/4	13/16
	75	20	6 3/8	13/16
	86	20 1/4	7	15/16
	96	20 1/4	7 1/4	15/16
S24	80	24	7	7/8
	90	24	7 1/8	7/8
	100	24	7 1/4	7/8
	106	24 1/2	7 7/8	1 1/16
	121	24 1/2	8	1 1/16

Wide Flange I-Beams

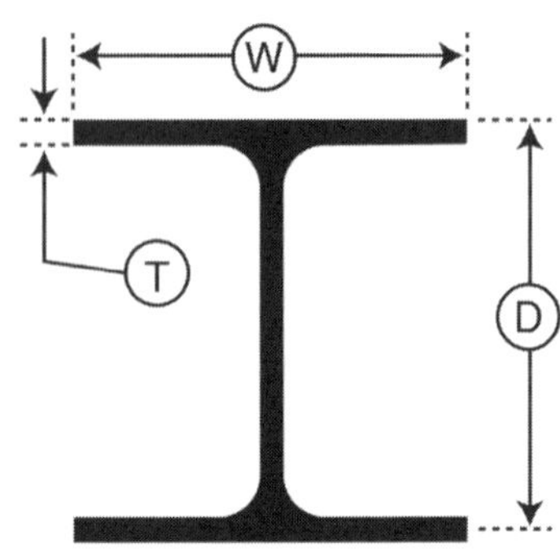
W
T
D

Section number	Wt. Per foot (lbs)	Dimensions		
		D	W	T
W8	10	7 7/8	4	3/16
	15	8 1/8	4	5/16
	18	8 1/8	5 1/4	5/16
	21	8 1/4	5 1/4	3/8
	24	7 7/8	6 1/2	3/8
	31	8	8	7/16
W10	12	9 7/8	4	3/16
	22	10 1/8	5 3/4	3/8
	26	10 3/8	5 3/4	7/16
	30	10 1/2	5 3/4	1/2
	33	9 3/4	8	7/16
	49	10	10	9/16

Data supplied for reference only. Dimensions and weights may be approximated, contanct steel supplier for specific details.

Wide Flange I-Beams *Continued*

Section number	Wt. Per foot (lbs)	Dimensions		
		D	W	T
W12	14	11 7/8	4	1/4
	22	12 1/4	4	7/16
	26	12 1/4	6 1/2	3/8
	35	12 1/2	6 1/2	1/2
	40	12	8	1/2
	45	12	8	9/16
	53	12	10	9/16
	65	12 1/8	12	5/8
W14	22	13 3/4	5	5/16
	26	13 7/8	5	7/16
	30	13 7/8	6 3/4	3/8
	38	14 1/8	6 3/4	1/2
	43	13 5/8	8	1/2
	48	13 3/4	8	5/8
	53	13 7/8	8	11/16
	61	13 7/8	10	5/8
	82	14 1/4	10 1/8	7/8
	90	14	14 1/2	11/16

Section number	Wt. Per foot (lbs)	Dimensions D	W	T
W16	26	15 3/4	5 1/2	3/8
	36	15 7/8	7	7/16
	45	16 1/8	7	9/16
	50	16 1/4	7 1/8	5/8
	57	16 3/8	7 1/8	11/16
	67	16 3/8	10 1/4	11/16
	77	16 1/2	10 1/4	3/4
	89	16 3/4	10 3/8	7/8
	100	17	10 3/8	1
W18	35	17 3/4	6	7/16
	40	17 7/8	6	1/2
	50	18	7 1/2	9/16
	60	18 1/4	7 1/2	11/16
	65	18 3/8	7 5/8	3/4
	71	18 1/2	7 5/8	13/16
	76	18 1/4	11	11/16
W21	44	20 5/8	6 1/2	7/16
	62	21	8 1/4	5/8
	73	21 1/4	8 1/4	3/4
	83	21 3/8	8 3/8	13/16
	93	21 5/8	8 3/8	15/16
	101	21 3/8	12 1/4	13/16

Wide Flange I-Beams

Continued

W
T
D

Section number	Wt. Per foot (lbs)	Dimensions		
		D	W	T
W24	55	23 5/8	7	1/2
	68	23 3/4	9	9/16
	84	24 1/8	9	3/4
	94	24 1/4	9 1/8	7/8
	104	24	12 3/4	3/4
	131	24 1/2	12 7/8	15/16
	146	24 3/4	12 7/8	1 1/16
	162	25	13	1 1/4
W27	84	26 3/4	10	5/8
	94	26 7/8	10	3/4
	102	27 1/8	10	13/16
	114	27 1/4	10 1/8	15/16
	146	27 3/8	14	1
	161	27 5/8	14	1 1/16
W30	124	30 1/8	10 1/2	15/16
	132	30 1/4	10 1/2	1
	148	30 5/8	10 1/2	1 3/16

Weight Estimator

Estimated weight of single part wire rope slings with pressed mechanical type sleeve connectors.

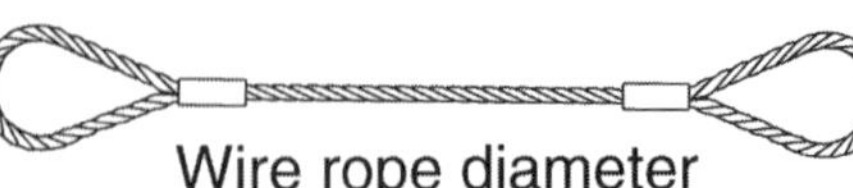

Wire rope diameter

Sling Length in feet	3/4"	1"	1-1/2"	2"	2-1/2"	3"	3-1/2"	4"
10	16	30	78	156				
15	22	40	98	193	342			
20	27	49	119	230	400	600		
25	32	58	140	267	458	683	1,024	
30	37	67	161	304	516	766	1,137	1,621
35	42	77	182	341	574	849	1,251	1,769
40	48	86	202	378	632	932	1,364	1,917
45	53	95	223	415	690	1,015	1,478	2,065
50	58	104	244	452	748	1,098	1,591	2,213
55	63	114	265	489	806	1,181	1,705	2,361
60	68	123	286	526	864	1,264	1,818	2,509
65	74	132	306	563	922	1,347	1,932	2,657
70	79	141	327	600	980	1,430	2,045	2,805
75	84	151	348	637	1,038	1,513	2,159	2,953
80	89	160	369	674	1,096	1,596	2,272	3,101
85	94	169	390	711	1,154	1,679	2,386	3,249
90	100	178	410	748	1,212	1,762	2,499	3,397
95	105	188	431	785	1,270	1,845	2,613	3,545
100	110	197	452	821	1,328	1,928	2,726	3,693

weights shown are in lbs

Shackle Weight Estimator

Size	Estimated weights – contact manufacturer for exact weight of each size shackle.	Weight
3/4"		3 lbs
7/8"		4 lbs
1"		6 lbs
1-1/4"		12 lbs
1-1/2"		21 lbs
1-3/4"		34 lbs
2"		53 lbs
2-1/2"		99 lbs
3"		154 lbs
3-1/2"		265 lbs
4"		338 lbs

STANDARDS

Disclaimer

The information contained in this publication was obtained from sources believed to be reliable at the time this publication was written. The publisher and author make no representation or guarantee as to the correctness or sufficiency of any information contained herein, nor a guarantee of results based upon the use of this information, and disclaims all warranties whether implied, express or statutory, including without limitation, implied warranties of merchantability, fitness for use and fitness for a particular purpose. You assume the entire risk as to the use of this information, and The publisher and the author assumes no liability in connection with either the information presented or use of the suggestions made in this publication.

Hardware Inspection

Per ASME B30.26

Initial Inspection:
Prior to use, all new rigging hardware shall be inspected. Written records are not required.

Frequent Inspection:
A visual inspection shall be performed each day before the rigging hardware is used. Written records are not required.

Periodic Inspection:
A complete inspection of the rigging hardware shall be performed by a designated person at an interval not to exceed one year. The frequency is be based on; the frequency of use, service conditions, and other experience gained. Written records are not required.

Removal Criteria
Rigging hardware shall be removed from service if damage such as the following is visible:

- missing or illegible identification or markings
- heat damage, weld spatter, arc strikes, excessive pitting, corrosion, nicks or gouges
- bent, twisted, distorted, stretched, elongated,cracked, or broken load-bearing components
- reduction (stretching) at any point around the body or pin, incomplete pin engagement, or excessive thread damage
- or any other conditions that cause doubt as to the continued use of the rigging hardware

Requirements shown are abridged and DO NOT address all the specific requirements for each type of rigging hardware. Refer to ASME B30.26 for a complete listing of these requirements.

Refer to ASME B30.26 for additional information. ASME Consensus Standards are usually more rigorous than state and federal OSHA requirements, compliance is voluntary unless otherwise required.

General requirements for slings

Safe operating practices (OSHA 1910.184) - Whenever any sling is used the following practices shall be observed:

1. Slings that are damaged or defective shall not be used.
2. Slings shall not be shortened with knots or bolts or other makeshift devices.
3. Sling legs shall not be kinked.
4. Slings shall not be loaded in excess of their rated capacities.
5. Slings used in a basket hitch shall have the loads balanced to prevent slippage.
6. Slings shall be securely attached to their loads.
7. Slings shall be padded or protected from the sharp edges of their loads.
8. Suspended loads shall be kept clear of all obstructions.
9. All employees shall be kept clear of loads about the be lifted and of suspended loads.
10. Hands or fingers shall not be placed between the sling and its load while the sling is being tightened around the load.
11 Shock loading is prohibited.
12. A sling shall not be pulled from under a load when the load is resting on the sling.

Inspections - Each day before being used, the sling and all fastenings and attachments shall be inspected for damage or defects by a competent person designated by the employer. Additional inspections shall be performed during sling use where service conditions warrant. Damaged or defective slings shall be immediately removed from service.

Wire rope inspection chart

The inspection criteria for wire rope depends entirely on the intended use, and the applicable regulatory code that the inspection is based on. The following table lists some of the more common standards that address inspection criteria.

Rope Replacement based on the number of Broken WIres		Number of Broken wires			
		Running Ropes		Standing Ropes	
		In one:	In one:	In one:	At end conne-ction
Standard	Equipment	lay	strand	lay	
ASME B30.2	Overhead & Gantry Cranes	12	4	(not specified)	
ASME B30.3	Hammerhead Tower Cranes	6	4	3	2
ASME B30.4	Portal, Tower, PillarCranes	6	3	3	2
ASME B30.5	Crawler, locomotive, Truck cranes	6	3	3	2
ASME B30.6	Derricks	6	3	3	2
ASME B30.7	Base Mounted Drum Hoists	6	3	3	2
ASME B30.8	Floating Cranes and Derricks	6	3	3	2
ASME B30.9	SLINGS	10	5	(not specified)	

This information is provided as an overview of the subject matter covered. The user of this book must understand that the publisher is not engaged in rendering legal, engineering, or other professional services. If legal or other expert assistance is required, the services of a certified professional person should be sought.

Alloy Steel Chain Slings(ASME B30.9) - An alloy steel chain sling shall be removed from service if conditions such as the following are present:

1. Missing or illegible sling identification.
2. Cracks or breaks
3. Excessive wear, nicks, or gouges.
4. Stretched chain links or components
5. Bent, twisted, or deformed chain links or components.
6. Evidence of hear damage.
7. Excessive pitting or corrosion.
8. Lack of ability of chain or components to hinge (articulate) freely.
9. Weld splatter.
10. For hooks, removal criteria as stated in ASME B30.10
11. Other conditions, including visible damage, that cause double as to the continued use of the sling.

OSHA Requirements

Alloy steel chain slings - OSHA 1910.184

Sling identification.

- Alloy steel chain slings shall have permanently affixed durable identification stating size, grade, rated capacity, and reach.
- Makeshift links or fasteners formed from bolts or rods, or other such attachments, shall not be used.

Safe operating temperatures.

- Alloy steel chain slings shall be permanently removed from service if they are heated above 1000 deg. F. When exposed to service temperatures in excess of 600 deg. F, maximum working load limits shall be reduced in accordance with the manufacturer's recommendations.

Inspections

- Each day before being used, the sling and all fastenings and attachments shall be inspected for damage or defects by a competent person designated by the employer.
- A thorough periodic inspection of alloy steel chain slings in use shall be made on a regular basis.
- The employer shall make and maintain a record of the most recent month in which each alloy steel chain sling was thoroughly inspected, and shall make such record available for examination.
- The thorough inspection of alloy steel chain slings shall be performed by a competent person designated by the employer, and shall include a thorough inspection for wear, defective welds, deformation and increase in length. Where such defects or deterioration are present, the sling shall be immediately removed from service.

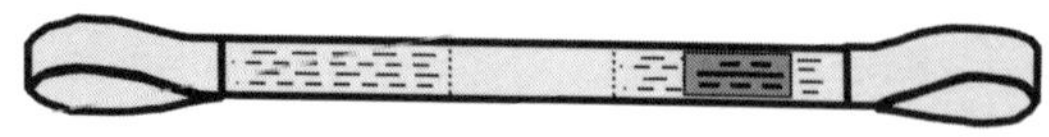

Nylon Web Slings (ASME B30.9) - A synthetic webbing sling shall be removed from service if conditions such as the following are present:

1. Missing or illegible sling identification.
2. Acid or caustic burns.
3. Melting or charring of any part of the sling.
4. Holes, tears, cuts, or snags.
5. Broken or worn stitching in load bearing splices.
6. Excessive abrasive wear.
7. Knots in any part of the sling.
8. Discoloration and brittle or stiff areas on any part of the sling, which may mean chemical or ultraviolet/sunlight damage.
9. Fitting that are pitted, corroded, cracked, bent, twisted, gouged, or broken.
10. For hooks, removal criteria as stated in ASME B30.10
11. Other conditions, including visible damage, that cause double as to the continued use of the sling.

OSHA Requirements

Synthetic web slings - OSHA 1910.184

Sling identification.

- Each sling shall be marked or coded to show the rated capacities for each type of hitch and type of synthetic web material.

Safe operating temperatures.

- Synthetic web slings of polyester and nylon shall not be used at temperatures in excess of 180 deg. F. Polypropylene web slings shall not be used at temperatures in excess of 200 deg. F.

Removal from service.

Synthetic web slings shall be immediately removed from service if any of the following conditions are present:

- Acid or caustic burns
- Melting or charring of any part of the sling surface
- Snags, punctures, tears or cuts
- Broken or worn stitches
- Distortion of fittings

Environmental conditions.

When synthetic web slings are used, the following precautions shall be taken:

- Nylon web slings shall not be used where fumes, vapors, sprays, mists or liquids of acids or phenolics are present.
- Polyester and polypropylene web slings shall not be used where fumes, vapors, sprays, mists or liquids of caustics are present.
- Web slings with aluminum fittings shall not be used where fumes, vapors, sprays, mists or liquids of caustics are present.

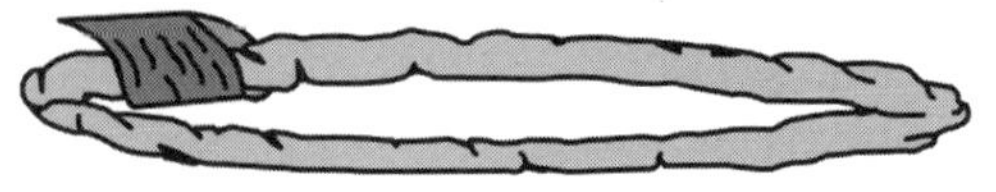

Polyester Round Slings (ASME B30.9) - A synthetic round sling shall be removed from service if conditions such as the following are present:

1.Missing or illegible sling identification.
2. Acid or caustic burns.
3. Evidence of heat damage.
4. Holes, tears, cuts, abrasive wear, or snags that expose the core yarns.
5. Broken or damaged core yarns.
6. Weld splatter that exposes core yarns.
7. Round slings that are knotted.
8. Discoloration and brittle or stiff areas on any part of the slings, which may mean chemical or ultraviolet/sunlight damage.
9. Fitting that are pitted, corroded, cracked, bent twisted, gouged, or broken.
10. For hooks, removal criteria as stated in ASME B30.10
11. Other conditions, including visible damage, that cause double as to the continued use of the sling.

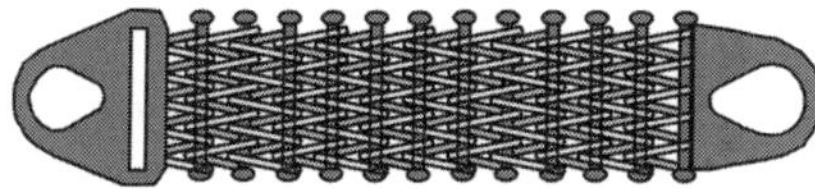

Wire Mesh Slings (ASME B30.9) - A metal mesh sling shall be removed from service if conditions such as the following are present:

1. Missing or illegible sling identification.
2. Broken weld or a broken brazed joint along the sling edge
3. Broken wire in any part of the mesh.
4. Reduction in wire diameter of 25% due to abrasion or 15% due to corrosion.
5. Lack of flexibility due to distortion of the mesh.
6. Distortion of the choker fitting so the depth of the slot is increased by more that 10%
7. Distortion of either end fitting so the width of the eye opening is decreased by more than 10%
8. A 15% reduction of the original cross-sectional area of any point around the hook opening of the end fitting.
9. Visible distortion of either end fitting out of its plane.
10. Cracked end fitting.
11. Slings in which the spirals are locked or without free articulation shall not be used.
12. Fitting that are pitted, corroded, cracked, bent, twisted, gouged, or broken.
13. Other conditions, including visible damage, that cause doubt as to the continued use of the sling.

For additional information, please refer to the OSHA and ASME standards.

Wire Rope Slings (ASME B30.9) - A wire rope sling shall be removed from service if conditions such as the following are present:

1. Missing or illegible sling identification.
2. Broken Wires:
   - For strand-laid and single-part slings, ten randomly distributed broken wires in on rope lay, or five broken wires in one strand in one rope lay.
   - For cable-laid slings, 20 broken wires per lay.
   - For six-part braided slings, 20 broken wires per braid
   - For eight-part braided slings, 40 broken wires per braid.
3. Severe localized abrasion or scraping.
4. Kinking, crushing, bird caging, or any other damage resulting in damage to the rope structure.
5. Evidence of heat damage
6. End attachments that are cracked, deformed, or worn to the extent that the strength of the sling is substantially affected.
7. Severe corrosion of the rope, end attachments, or fittings.
8. for hooks, removal criteria at stated in ASME B30.10.
9. Other conditions, including visible damage, that cause doubt as to the continued use of the sling.

Wire rope slings - OSHA 1910.184

OSHA Requirements

Safe operating temperatures.

- Fiber core wire rope slings of all grades shall be permanently removed from service if they are exposed to temperatures in excess of 200 deg. F. When nonfiber core wire rope slings of any grade are used at temperatures above 400 deg. F or below minus 60 deg. F, recommendations of the sling manufacturer regarding use at that temperature shall be followed.

Removal from service.

Wire rope slings shall be immediately removed from service if any of the following conditions are present:

- Ten randomly distributed broken wires in one rope lay, or five broken wires in one strand in one rope lay.
- Wear or scraping of one-third the original diameter of outside individual wires.
- Kinking, crushing, bird caging or any other damage resulting in distortion of the wire rope structure.
- Evidence of heat damage.
- End attachments that are cracked, deformed or worn.
- Hooks that have been opened more than 15 percent of the normal throat opening measured at the narrowest point or twisted more than 10 degrees from the plane of the unbent hook.
- Corrosion of the rope or end attachments.

CRANE REQUIREMENTS WITH MANBASKETS

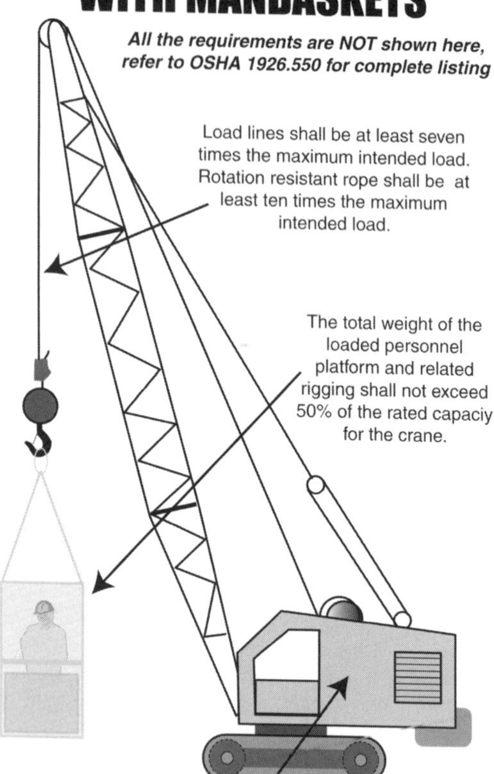
All the requirements are NOT shown here, refer to OSHA 1926.550 for complete listing
Load lines shall be at least seven times the maximum intended load. Rotation resistant rope shall be at least ten times the maximum intended load.
The total weight of the loaded personnel platform and related rigging shall not exceed 50% of the rated capaciy for the crane.

All brakes and locking devices shall be engaged when the occupied personnel platform is in a stationary working position.

Cranes that have live booms are prohibited !

This information provides a generic, non-exhaustive overview of the OSHA standard on suspended personnel platforms. Standards and interpretations change over time, you should always check current OSHA compliance requirements for your specific requirements.

29 CFR 1926.550 addresses the use of personnel hoisting in the construction industry, and 29 CFR 1910.180 addresses the use of personnel hoisting in general industry

SUSPENDED PERSONNEL PLATFORMS

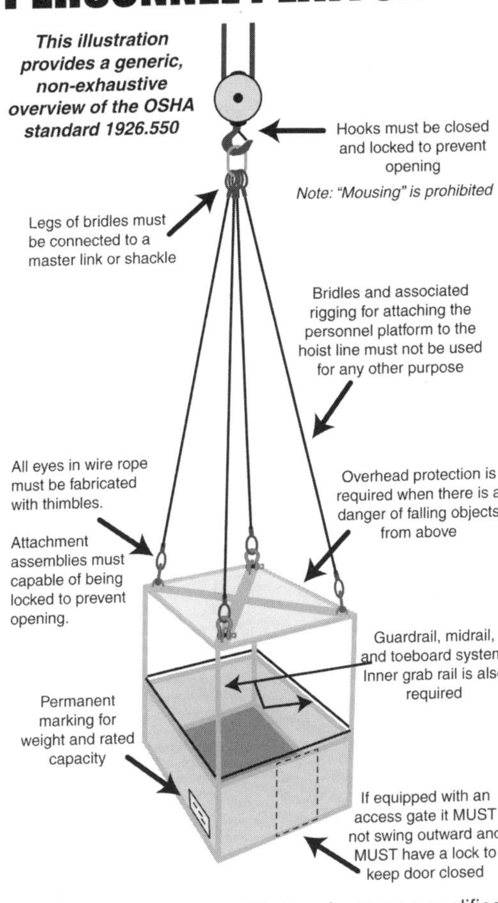
This illustration provides a generic, non-exhaustive overview of the OSHA standard 1926.550
Hooks must be closed and locked to prevent opening
Note: "Mousing" is prohibited
Legs of bridles must be connected to a master link or shackle
Bridles and associated rigging for attaching the personnel platform to the hoist line must not be used for any other purpose
All eyes in wire rope must be fabricated with thimbles.
Attachment assemblies must capable of being locked to prevent opening.
Overhead protection is required when there is a danger of falling objects from above
Guardrail, midrail, and toeboard system. Inner grab rail is also required
Permanent marking for weight and rated capacity
If equipped with an access gate it MUST not swing outward and MUST have a lock to keep door closed

Must be designed by a qualified engineer, or a qualified person competent in structural design, and fabricated and welded by a qualified welder.

The OSHA rules on crane suspended personnel platforms contain many specifics that are not covered in this book. Refer to 29 CFR Part 1926.550 for the current OSHA compliance requirements.

API RP 2D Overview

A QUALIFIED OPERATOR is a person so designated by the employer. Who has appropriate offshore experience and training, documented training in both classroom and "hands on" is required. There are two levels of proficiency, non-mechanical and mechanical. The qualified operator should be qualified to perform pre-use and monthly inspection. A qualified operator is also considered a qualified rigger.

THE QUALIFIED OPERATOR WILL:

- Be responsible for the operations under direct control
- Stop and refuse to handle loads as safety dictates
- Be a ware of crane's operating characteristics
- Be familiar with the equipment and its care
- Never start machine movement unless load or signal person is within range of vision
- Respond to signals only from appointed signal person
- Obey an emergency stop signal given by any person a t any time
- Verify that appropriate load charts are in place for the crane configuration in use
- Secure the crane against swinging when not in use
- Will not operate crane in proximity of helidecks while helicopter is landing or taking off
- Insure there is sufficient lighting when cranes used at night
- Will not allow field welding on load hooks or sling hooks
- The operator will maintain log of pre use inspection.

Qualified Rigger

A qualified rigger is a person with training and experience that has successful y completed a rigger training program.

A rigger is anyone who attaches or detaches lifting equipment to loads or lifting devices.

Qualified Inspector

A qualified inspector is a person so designated by the employer who has appropriate experience and training. In addition to meeting the requirements of a qualified operator, has attended formal training on crane maintenance and troubleshooting. Documented training, both classroom and "hands on" is required.

Offshore Rigging

API RP 2D requires before starting to lift, verify:

- If there is a slack rope condition, the rope is properly seated on the drum and in the sheaves.
- The correct slings have been selected for the weight to be lifted.
- The load is free to be lifted.
- Multiple part lines are not twisted around each other in such a manner that all of the lines will not separate upon application of load.
- The hook is brought over the load in such a manner to minimize swinging.

API RP 2D requires that during lifting, care should be taken that:

- Acceleration or deceleration of the moving load is accomplished in a smooth manner.
- That load, boom, or other parts of the machine do not contact any obstruction. The operator should engage the controls smoothly to avoid excessive stress on the crane.
- When rotating the crane, sudden starts and stops should be avoided.
- Rotational speed shall be such that the load does not swing out beyond the radius at which it can be controlled.

API RP 2D requires that during personnel transfer:

- All hooks used for support of personnel will have a safety latch. (latch with redundant locking)
- When making personnel lifts, the load will be under control in both up and down directions.
- All personnel to be lifted will use approved personnel flotation devices.
- Personnel riding on net type carriers should stand on outer rim facing inward.

Offshore Slings

API RP 2D requires that:

- Wire rope slings should not be field fabricated.
- If circumstances require field fabrication qualified inspectors should perform this function using accepted practices.
- Zinc or resin poured sockets shall be fabricated in accordance with API recommended practices
- Slings of all types will be proof loaded by the sling manufacturer per industry recommendations.
- All slings will be labeled with manufacturer and pertinent working load limits, proof test certification number, length, diameter, and date of proof test.
- Rated loads of a sling are different for each of the three basic methods of rigging (vertical, choker, basket).
- Rated loads of a sling is different for each of the methods of rigging based on construction of the rope, web material and width, etc.
- The rated loads will be indicated on the heavy duty tags attached to each type of sling at the time it is fabricated.
- Slings of other than wire rope construction will be used, inspected, and tested in accordance with the. Sling manufacturer and industry recommendations

Qualified Crane Operator

American Petroleum Institute (API) Recommended Practices for Operations and Maintenance of Offshore Cranes.

This information is provided as an overview of the subject matter covered. The user of this book must understand that the publisher is not engaged in rendering legal, engineering, or other professional services. If legal or other expert assistance is required, the services of a certified professional person should be sought.

Effective Date: 1/18/02

This new rule is intended to protect employees from the hazards associated with steel erection. Any employees involved in the construction, alteration and/or repair of single and multi-story buildings, bridges and other structures where steel erection occurs fall under this rule. It does NOT cover electrical transmission towers, communication and broadcast towers or tanks.

Steel erection activities include hoisting, connecting, welding, bolting and rigging structural steel, steel joists and metal buildings; installing metal deck, siding systems, miscellaneous metals, ornamental iron and similar materials; and moving point-to-point while performing these activities.

§1926.753 Hoisting and Rigging requires:

- A Pre-shift visual inspection of cranes by a competent person
- A certification that this pre-shift visual inspection was done shall be available
- The operator is responsible for those operations under the operator's direct control
- A qualified rigger shall inspect the rigging prior to each shift
- Cranes may hoist employees with a personnel platform meeting the requirements of .550(g)
- Working under loads, routes must be preplanned to assure no employee required to work under a suspended load except:
  - Connectors
  - Employees responsible for hooking/unhooking the load

- When working under suspended loads:
  - Routes shall be pre-planned
  - Materials shall be rigged to prevent unintentional displacement
  - Hooks shall have self-closing latches
- Multiple lift Rigging procedure can be used only if:
  - Maximum of 5 members
  - Only structural members
  - All employees have been trained
- Components shall be designed with a 5 to 1 safety factor
- Total load shall not exceed the rated capacity of the hoisting equipment or the rigging equipment
- Rigging assembly shall be rigged with the members attached at their center of gravity and maintained reasonably level; rigged from the top down; rigged at least 7 ft. apart.

Pertinent Definitions

Competent Person - means one who is capable of identifying existing and predictable hazards in the surroundings or working conditions which are unsanitary, hazardous or dangerous to employees and who has authorization to take prompt corrective measures to eliminate them.

Connector - means an employee who, working with hoisting equipment, is placing and connecting structural members and/or components.

Qualified person - means one who by possession of a recognized degree, certificate or professional standing, or who by extensive knowledge, training and experience has successfully demonstrated the ability to solve or resolve problems relating to the subject matter, the work or the project.

References

The following list of regulations and standards is not exhaustive, but is intended to give a general idea of the key points addressed with them.

The ANSI/ASME standards are the foundation for most rigging requirements and standards. These "voluntary" standards are often "rolled into" OSHA requirements.

Listed below are the volumes related to cranes and rigging:

B30.1 Jacks
B30.2 Overhead and Gantry Cranes
B30.3 Construction Tower Cranes
B30.4 Portal, Tower, and Pedestal Cranes
B30.5 Mobile and Locomotive Cranes
B30.6 Derricks
B30.7 Base Mounted Drum Hoists
B30.8 Floating Cranes and Floating Derricks
B30.9 Slings
B30.10 Hooks
B30.11 Monorails and Underhung Cranes
B30.12 Handling Loads Suspended From Rotorcraft
B30.13 Storage/Retrieval Machines
B30.14 Side Boom Tractors
B30.15 Mobile Hydraulic Cranes (withdrawn see B30.5)
B30.16 Overhead Hoists (Underhung)
B30.17 Overhead and Gantry Cranes
B30.18 Stacker Cranes
B30.19 Cableways
B30.20 Below-the-Hook Lifting Devices
B30.21 Manually Lever Operated Hoists
B30.22 Articulating Boom Cranes
B30.23 Personnel Lifting Systems
B30.24 Container Cranes*
B30.25 Scrap and Material Handlers
B30.26 Rigging Hardware
B30.27 Material Placement Systems
B30.28 Balance-Lifting Units*

*in the developmental stage

The following is a listing of various Industry Standards and where to obtain them:

Global Engineering Documents
15 Inverness Way East
Englewood CO 80112
303-397-7956
www.global.ihs.com
They carry most worldwide standards, including ANSI/ASME

OSHA
Occupational Safety & Health Administration
200 Constitution Avenue, NW
Washington, DC 20210
www.osha.gov

ASME International
Three Park Avenue
New York, NY 10016-5990
800-843-2763
www.asme.org

American Petroleum Institute
1220 L Street NW
Washington, DC 20005
Phone 202-682-8000
www.api.org

INDUSTRY ASSOCIATION REFERENCES

WIRE ROPE TECHNICAL BOARD
P.O. BOX 14921
SHAWNEE MISSION, KS 66285
They have the "WIRE ROPE USER'S MANUAL" and the "WIRE ROPE SLING USERS MANUAL"

WEB SLING AND TIE DOWN ASSOCIATION
710 E. OGDEN AVENUE. SUITE 600
NAPERVILLE. IL 60563
They reccommend standards for manufacturing and using synthetic web and round slings.

SLINGS HARDWARE FORMULAS EXAMPLES REFERENCE STANDARDS EXTRAS

CANADIAN REFERENCES

ONTARIO MINISTRY OF LABOUR.
OPERATIONS DIVISION
400 UNIVERSITY A VENUE
TORONTO, ONTARIO M7A 1T7

They have the "OCCUPATIONAL HEALTH AND SAFETY ACT & REGULATIONS FOR CONSTRUCTION PROJECTS for INDUSTRIAL ESTABLISHMENTS"

WORKERS' COMPENSATION BOARD OF BRITISH COLUMBIA
PO BOX 5350, STN TERMINAL
VANCOUVER BC V6B 5L5
They have the "OCCUPATIONAL HEALTH AND SAFETY REGULATIONS"

ACRP

EXTRAS

Wire Rope types

Wire Rope is made of steel wires laid together to form a strand. These strands are laid together to form a rope, usually around a central core of either fiber or wire.

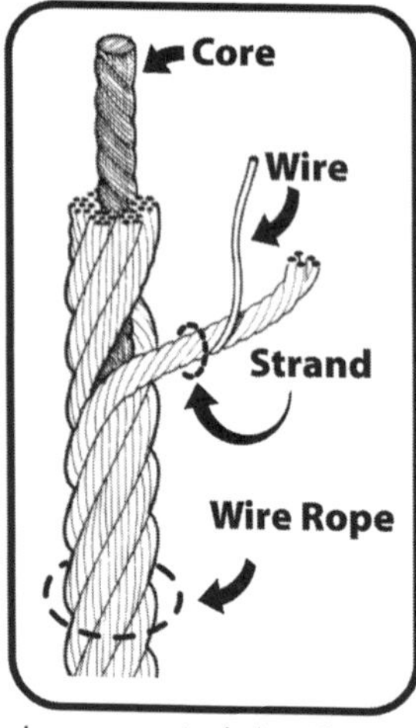
Core
Wire
Strand
Wire Rope

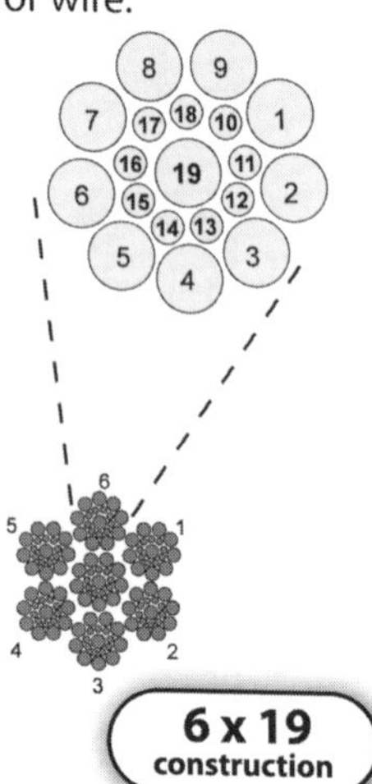
6 x 19
construction

In a numerical classification of rope construction, the first number is the number of strands; the second is the number is the number of wires. Thus, 6x19 means six strands of nineteen wires per strand.

FIBER CORE

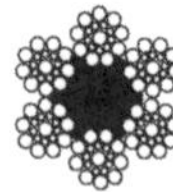

A fiber core is composed of a synthetic fiber such as polypropylene, or a natural fiber like jute or hemp.

IWRC

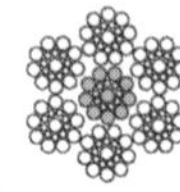

IWRC is the abbreviation for independent wire rope core. This wire core adds to the overall strength. This is the most common construction used today.

Wire Rope grades

Ropes are available in all 3 grades, Improved Plow Steel (IPS), Extra Improved Plow Steel (EIPS) and Extra Extra Improved Plow Steel (EEIPS).

Strand Classification

Strands are grouped according to the number of wires per strand and is related to the typical strength of the wire rope. The first number indicates the number of strands in the rope, the second number indicates the number of wires within each strand. The number of wires is nominal and can vary, as shown in the table below.

Classification	number of strands	number of wires
6 x 7	6	3 to 14
6 x 19	6	16 to 26
6 x 37	6	27 to 49
8 x 19	8	15 to 26

Wire Rope Lay Length

The lay length of a wire rope is the straight linear distance of one strand as it makes a complete revolution.

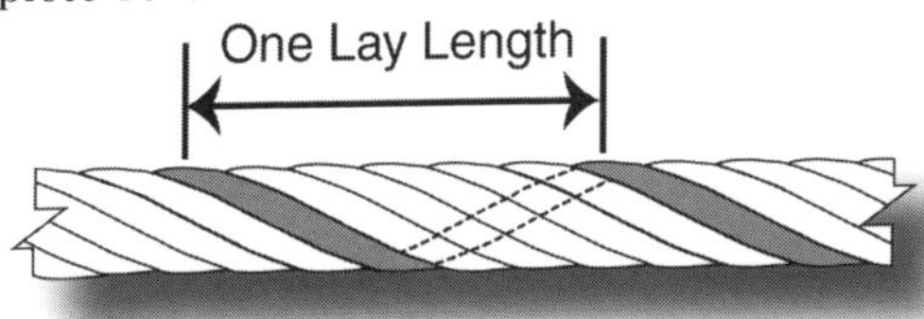
One Lay Length

Strength of wire rope

Breaking Strength "**The strength at which new, unused wire rope will break under a stationary load.**" As a wire rope wears over time, the breaking strength is naturally reduced.

Breaking strength should never be considered the wire rope's work load limit. The work load limit depends on application and the method used to create a finished assembly, and is always significantly lower than the breaking strength.

Measuring Rope Diameter

The correct diameter of a wire rope is the diameter of a circumscribed circle that will enclose all the strands. It's the largest cross-sectional measurement. You should make the measurement carefully with calipers.

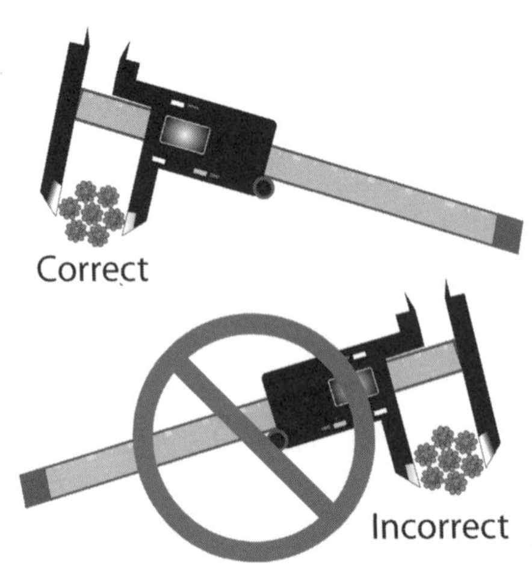
Correct
Incorrect

Wire Rope "Lay"

"Lay" has three meanings in rope design. The first two meanings are descriptive of the wire and strand positions in the rope. The third meaning is a length measurement used in manufacturing and inspection.

The direction strands lay in the rope -- right or left. When you look down a rope, strands of a right lay rope go away from you to the right. Left lay is the opposite. (It doesn't matter which direction you look.)

The relationship between the direction strands lay in the rope and the direction wires lay in the strands. In appearance, wires in regular lay appear to run straight down the length of the rope, and in lang lay, they appear to angle across the rope. In regular lay, wires are laid in the strand opposite the direction the strands lay in the rope. In lang lay, the wires are laid the same direction in the strand as the strands lay in the rope.

The length along the rope that a strand makes one complete spiral around the rope core. This is a measurement frequently used in wire rope inspection. Standards and regulations require removal when a certain number of broken wires per rope lay are found.

The lay of a rope affects its operational characteristics. Regular lay is more stable and more resistant to crushing than lang lay. While lang lay is more fatigue resistant and abrasion resistant, use is normally limited to single layer spooling and when the rope and load are restrained from rotation.

IWRC Breaking Strength

IWRC

Standard 6 x 19 and 6 x 37 classification ropes

Diameter (in.)	Approx. wt./ft. (lbs.)	Minimum Breaking Force		
		IPS	EIPS	EEIPS
3/16	--	--	--	--
1/4	0.116	2.94	3.4	--
5/16	0.18	4.58	5.27	--
3/8	0.26	6.56	7.55	8.3
7/16	0.35	8.89	10.2	11.2
1/2	0.46	11.5	13.3	14.6
9/16	0.59	14.5	16.8	18.5
5/8	0.72	17.9	20.6	22.7
3/4	1.04	25.6	29.4	32.4
7/8	1.42	34.6	39.8	43.8
1	1.85	44.9	51.7	56.9
1 1/8	2.34	56.5	65	71.5
1 1/4	2.89	69.4	79.9	87.9
1 3/8	3.5	83.5	96	106
1 1/2	4.16	98.9	114	125
1 5/8	4.88	115	132	146
1 3/4	5.67	133	153	169
1 7/8	6.5	152	174	192
2	7.39	172	198	217
2 1/8	8.35	192	221	244
2 1/4	9.36	215	247	272

Always consult the wire rope manufacturer for actual breaking strength, values shown are representative of typical wire rope and do not apply to all wire rope.

Breaking strength should never be considered the wire rope's work load limit. The work load limit depends on application and the method used to create a finished assembly, and is always significantly lower than the breaking strength.

Fiber Core Breaking Strength

FIBER CORE

Standard 6 x 19 and 6 x 37 classification ropes

Diameter (in.)	Approx. wt./ft. (lbs.)	Minimum Breaking Force	
		IPS	EIPS
3/16	0.059	1.55	1.71
1/4	0.105	2.74	3.02
5/16	0.164	4.26	4.69
3/8	0.0236	6.1	6.72
7/16	0.32	8.27	9.1
1/2	0.42	10.7	11.8
9/16	0.53	13.5	14.9
5/8	0.66	16.7	18.3
3/4	0.095	23.8	26.2
7/8	1.29	32.2	35.4
1	1.68	41.8	46
1 1/8	2.13	52.6	57.8
1 1/4	2.63	64.6	71.1
1 3/8	3.18	77.7	85.5
1 1/2	3.78	92	101
1 5/8	4.44	107	118
1 3/4	5.15	124	137
1 7/8	5.91	141	156
2	6.72	160	176
2 1/8	7.59	179	197
2 1/4	8.51	200	220

Breaking strength should never be considered the wire rope's work load limit. The work load limit depends on application and the method used to create a finished assembly, and is always significantly lower than the breaking strength.

Always consult the wire rope manufacturer for actual breaking strength, values shown are representative of typical wire rope and do not apply to all wire rope.

Strength of 7x7 aircraft cable

7 X 7 Strand Aircraft cable - Has high strength and rugged construction. Excellent for transmitting mechanical power. Commonly used for applications such as guying, control, and light hoisting or supporting.

7 X 7 Aircraft Cable		
Wire Rope Diameter	Breaking Strength, in lbs.	
	Galvanized	304 Stainless Steel
1/16"	480	480
3/32"	920	920
1/8"	1,700	1,700
5/32"	2,600	2,400
3/16"	3,700	3,700

Breaking strength should never be considered the wire rope's work load limit. The work load limit depends on application and the method used to create a finished assembly, and is always significantly lower than the breaking strength.

Always consult the wire rope manufacturer for actual breaking strength. values shown are representative of typical wire rope and do not apply to all wire rope.

Strength of 7x19 aircraft cable

7 X 19 Strand Aircraft cable - Has good strength and resistance to crushing loads. More flexible than the 7 X 7 construction.

7 X 19 Aircraft Cable		
Wire Rope Diameter	Breaking Strength, in lbs.	
	Galvanized	304 Stainless Steel
3/32"	1,000	920
1/8"	2,000	1,760
5/32"	2,800	2,400
3/16"	4,200	3,700
1/4"	7,000	6,400
5/16"	9,800	9,000
3/8"	14,400	12,000

Breaking strength should never be considered the wire rope's work load limit. The work load limit depends on application and the method used to create a finished assembly, and is always significantly lower than the breaking strength.

Always consult the wire rope manufacturer for actual breaking strength, values shown are representative of typical wire rope and do not apply to all wire rope.

Installing wire rope

When loading the drum, extreme care must be taken to ensure the drum is properly loaded, and each wrap is wound tightly against the preceding wrap.

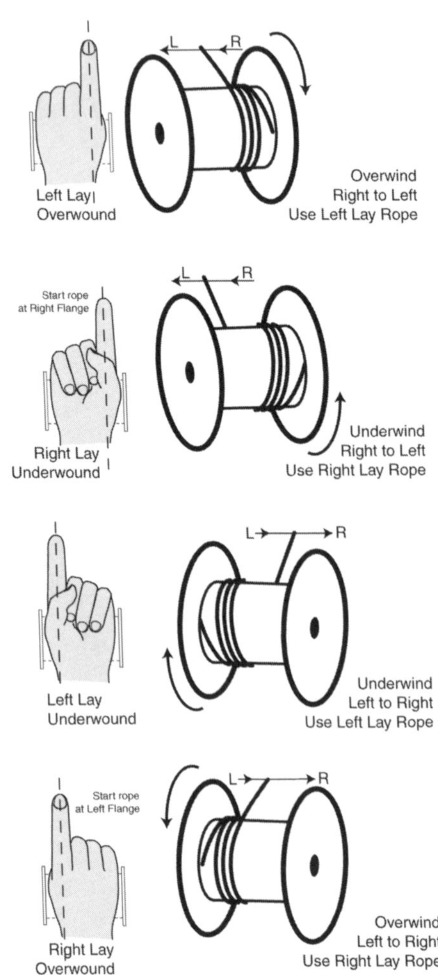
L
R
Left Lay
Overwound
Overwind
Right to Left
Use Left Lay Rope
L
R
Start rope
at Right Flange
Right Lay
Underwound
Underwind
Right to Left
Use Right Lay Rope
L
R
Left Lay
Underwound
Underwind
Left to Right
Use Left Lay Rope
L
R
Start rope
at Left Flange
Right Lay
Overwound
Overwind
Left to Right
Use Right Lay Rope

When installing new wire rope from a reel it is important that the rope be wound top-to-top or bottom-to-bottom. The illustration below shows how to properly load the new spool while keeping tension on the reel.

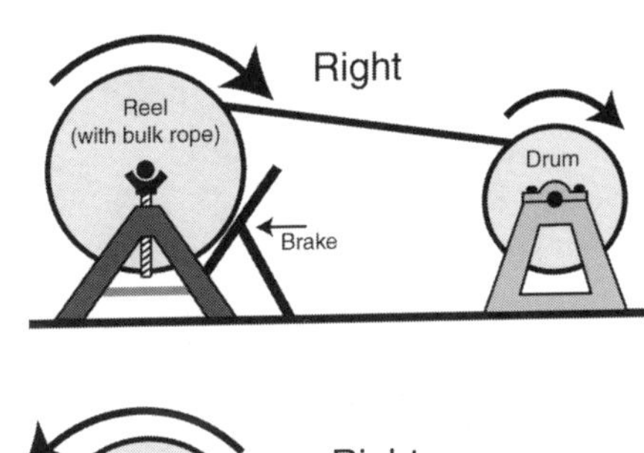
Right
Reel
(with bulk rope)
Drum
Brake

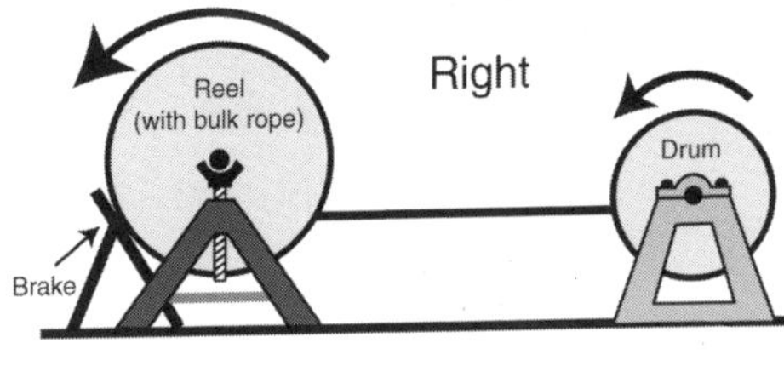
Right
Reel
(with bulk rope)
Drum
Brake

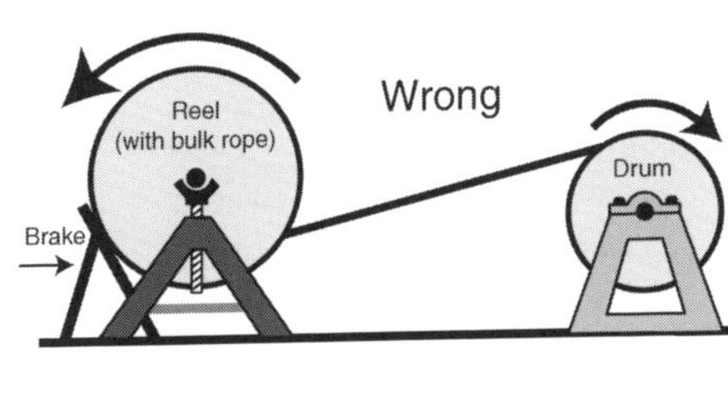
Wrong
Reel
(with bulk rope)
Drum
Brake

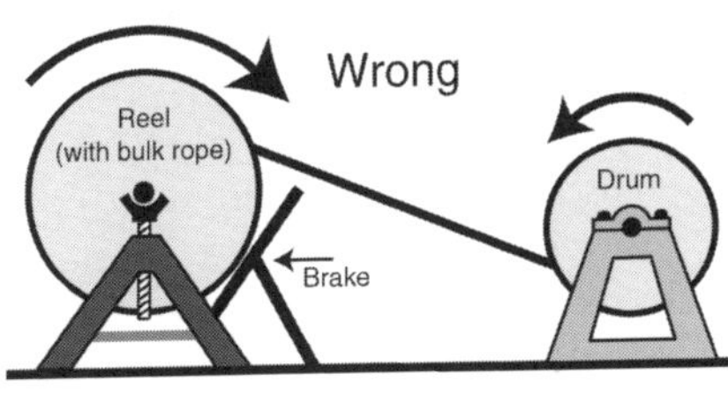
Wrong
Reel
(with bulk rope)
Drum
Brake

Hand Signals

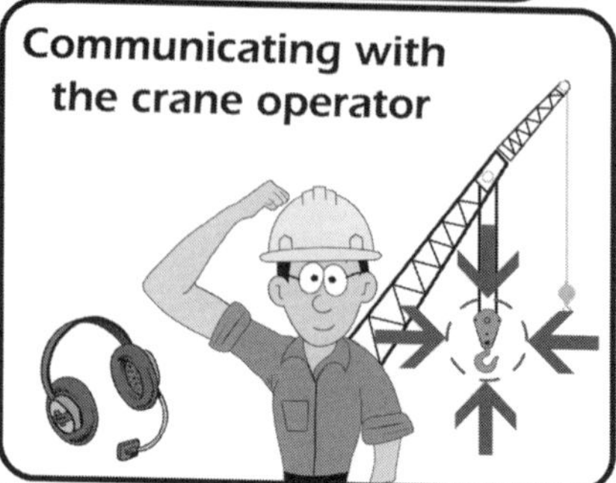
Communicating with
the crane operator

The signaler must always:

- Be in clear view of the crane operator
- Have a clear view of the load and the equipment at all times
- Keep people outside the load travel path
- Ensure the load does not pass above people
- Keep the crane away from power lines
- Watch for other overhead hazards that could endanger the load or people nearby

Duties of the Signalman

- Only one person shall be designated as a signalman.
- The signalman AND operator must be familiar with the Hand Signals being used.
- The signalman must be able to observe the load and other workmen at all times.
- The signalman must always be in plain view of the crane operator.

Multiple Signalers

- There should be only one designated signaler at a time - more than one will only confuse the operator.
- If signalers are changing frequently, the one in charge should be clearly visible as the person with signaling authority
- Wearing a bright vest, or different colored hard hat will help the crane operator quickly identify who is currently in charge of signaling.

Radio Use

Radios in place of hand signals

The use of radios is gaining popularity in today's workplace. Reliable voice activated headsets are preferred because they keep hands free for both the crane operator and signaler.

Cautions regarding radio use may include the following:

- Awareness of any explosive devices in general area *(radio transmissions have been known to cause premature detonation of explosives that use electric detonators)*
- Other electronics (potential for interference)
- Other radios nearby operating on the same frequency.

Some basics when using radio commands:

- Discuss lift plan with operator
- Use a secure frequency, free of distracting chatter
- Use first names not titles
  i.e.: "Jim" as opposed to "crane operator"

Commands names should be same as the hand signal names:

- Use whipline
- Boom down, hold load
- Boom up
- Etc..

- Once lift has begun the signaler should never break communication with the operator - this is referred to as "constant communication"
- Never unkey the mic while the load is moving. The signaler should repeat the command to let the operator know everything is alright:
  Slowly down, slow,slow, ...

- If signaler breaks communications (unkeys mic.) the operator should stop immediately.
- The crane operator should only continue when regular communications with signaler is re-established.
- If the operator needs to talk with the signaler, he should stop and give 1 blast of the horn to alert the signaler.
- The signaler can then unkey the mic so the crane operator can talk with him.

Hand Signals

Mobile Crane

Hand Signals

HOIST LOAD
SLOWLY RAISE LOAD
STOP

LOWER LOAD

SLOWLY LOWER LOAD
ALL STOP
EMERGENCY STOP

2 Hand
EXTEND BOOM
1 Hand

USE MAIN HOIST

2 Hand
RETRACT BOOM
1 Hand

DOG EVERYTHING
SWING BOOM

USE WHIP LINE

Refer to ASME B30.5 and OSHA 1926.550 for additional information

These hand signals were created using ASME B30.5 as the guideline and basis of development.

Some special operations may require adaptations of these basic signals.

Hand Signals

RAISE BOOM
RAISE BOOM
& LOWER LOAD
LOWER BOOM
LOWER BOOM
& RAISE LOAD

SWING

Arm extended, point with finger the direction of swing of boom.

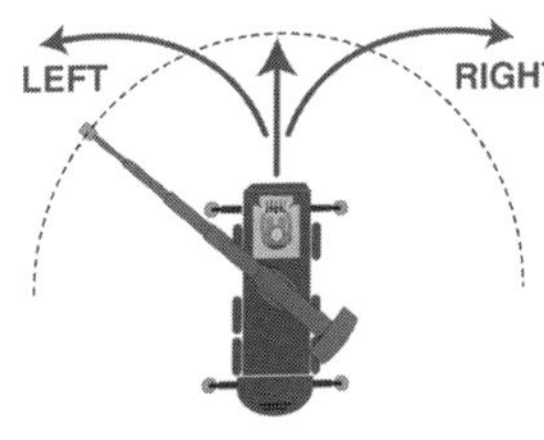
LEFT
RIGHT

Voice Commands when using a radio:

- When giving the "Swing" command:
- Give directions to the operators right or left
- Give approximate length of swing

Hand Signals

OVERHEAD CRANES

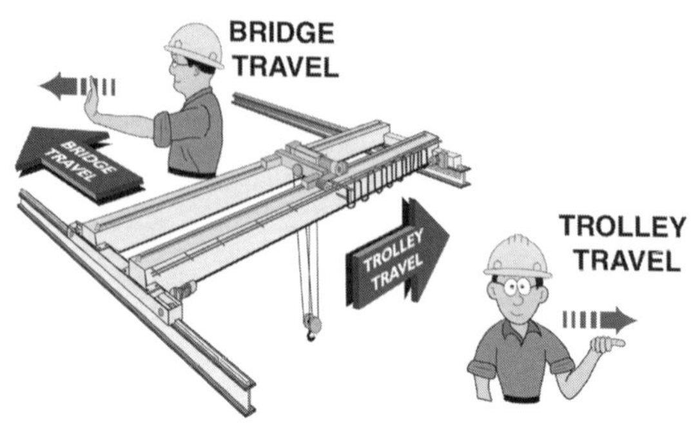

BRIDGE TRAVEL
BRIDGE TRAVEL
TROLLEY TRAVEL
TROLLEY TRAVEL

Signals adapted from ASME B30.2-2005 (Overhead and Gantry Cranes)

Special operations may require additions or modifications of these signals.

MAGNET IS DISCONNECTED

Crane Operator ONLY
Spread both hands apart, palms up.

1
2

MULTIPLE TROLLEYS

Hold up ONE finger for the block marked "1" and TWO fingers for the block Marked "2". Then follow with regular hand signals.

Hand Signals

STOP
Right arm extended ,palm down and open, move arm back and forth horizontally

ALL STOP!

EMERGENCY STOP

An emergency stop signal must be accepted from any person.

It is important that the crane operator immediately react to this signal, because the person giving this signal may recognize a potentially hazardous situation that the crane operator or signalman is not aware of.

FRACTION/DECIMAL/MILLIMETER CONVERSION CHART

fraction	decimal	mm	fraction	decimal	mm
1/64	0.0156	0.3969	33/64	0.5156	13.0969
1/32	0.0313	0.7938	17/32	0.5313	13.4938
3/64	0.0469	1.1906	35/64	0.5469	13.8906
1/16	0.0625	1.5875	39341	0.5625	14.2875
5/64	0.0781	1.9844	37/64	0.5781	14.6844
3/32	0.0938	2.3813	19/32	0.5938	15.0813
7/64	0.1094	2.7781	19/32	0.6094	15.4781
1/8	**0.125**	**3.175**	**5/8**	**0.625**	**15.875**
9/64	0.1406	3.5719	41/64	0.6406	16.2719
5/32	0.1563	3.9688	21/32	0.6563	16.6688
11/64	0.1719	4.3656	43/64	0.6719	17.0656
3/16	0.1875	4.7625	11/16	0.6875	17.4625
13/64	0.2031	5.1594	45/64	0.7031	17.8594
7/32	0.2188	5.5563	23/32	0.7188	18.2563
15/64	0.2344	5.9531	47/64	0.7344	18.6531
1/4	**0.25**	**6.35**	**3/4**	**0.75**	**19.05**
17/64	0.2656	6.7469	49/64	0.7656	19.4469
9/32	0.2813	7.1438	25/32	0.7813	19.8438
19/64	0.2969	7.5406	51/64	0.7969	20.2406
5/16	0.3125	7.9375	13/16	0.8125	20.6375
21/64	0.3281	8.3344	53/64	0.8281	21.0344
11/32	0.3438	8.7313	27/32	0.8438	21.4313
23/64	0.3594	9.1281	55/64	0.8594	21.8281
3/8	**0.375**	**9.525**	**7/8**	**0.875**	**22.225**
25/64	0.3906	9.9219	57/64	0.8906	22.6219
13/32	0.4063	10.3188	29/32	0.9063	23.0188
27/64	0.4219	10.7156	59/64	0.9219	23.4156
7/16	0.4375	11.1125	15/16	0.9375	23.8125
29/64	0.4531	11.5094	61/64	0.9531	24.2094
15/32	0.4688	11.9063	31/32	0.9688	24.6063
31/64	0.4844	12.3031	63/64	0.9844	25.0031
1/2	**0.5**	**12.7**	**1**	**1**	**25.4**

Multiplication Table

X	0	1	2	3	4	5	6	7	8	9	10	11	12
0	0	0	0	0	0	0	0	0	0	0	0	0	0
1	0	1	2	3	4	5	6	7	8	9	10	11	12
2	0	2	4	6	8	10	12	14	16	18	20	22	24
3	0	3	6	9	12	15	18	21	24	27	30	33	36
4	0	4	8	12	16	20	24	28	32	36	40	44	48
5	0	5	10	15	20	25	30	35	40	45	50	55	60
6	0	6	12	18	24	30	36	42	48	54	60	66	72
7	0	7	14	21	28	35	42	49	56	63	70	77	84
8	0	8	16	24	32	40	48	56	64	72	80	88	96
9	0	9	18	27	36	45	54	63	72	81	90	99	108
10	0	10	20	30	40	50	60	70	80	90	100	110	120
11	0	11	22	33	44	55	66	77	88	99	110	121	132
12	0	12	24	36	48	60	72	84	96	108	120	132	144

Phonetic Alphabet

Letter	Pronunciation
A	Alpha (AL fah)
B	Bravo (BRAH VOH)
C	Charlie (CHAR lee)
D	Delta (DELL tah)
E	Echo (ECK oh)
F	Foxtrot (FOKS trot)
G	Golf (GOLF)
H	Hotel (hoh TELL)
I	India (IN dee ah)
J	Juliett (JEW lee ETT)
K	Kilo (KEY loh)
L	Lima (LEE mah)
M	Mike (MIKE)
N	November (no VEM ber)
O	Oscar (OSS cah)
P	Papa (pah PAH)
Q	Quebec (keh BECK)
R	Romeo (ROW me oh)
S	Sierra (see AIR rah)
T	Tango (TANG go)
U	Uniform (YOU nee form)
V	Victor (VIK tah)
W	Whiskey (WISS key)
X	X Ray (ECKS RAY)
Y	Yankee (YANG key)
Z	Zulu (ZOO loo)

Note: The syllables printed in capital letters are to be stressed

Changes made in this printing

We try to be diligent when proofing the Rigging Handbook but we still overlook some mistakes – sorry. When we find a mistake or typo we correct these prior to a new print run, and sometimes update pages to provide better illustrations at the request of our readers.

Changes made in the April 2008 printing:

Page	Change/revisions
19	UPDATED - Clarified cautions and warnings about using the capacity table values vs actual WLL
20	TYPO - Changed HS to HT (to represent Hand Tucked)
72	Combined previous content with page 71, added WLL cautions
74	UPDATED - Revised and included alternate methods (turned tail) for securing wire rope ends on wedge sockets
87	Clarified the use of wire rope clips for rigging
88	GRAPHIC ILLUSTRATION ERROR - Corrected illustration to show the wire rope clips properly installed
91	NEW - Added carbon steel shackle WLL's, revised illustration to incorporate both Carbon and Alloy
92	Clarified sling side loading with new examples
131	UPDATED - Modified formulas for inclined planes (old formula was conservative) to provide better results
139	Revised and expanded illustration for the eyebolts used with angular lifting
140	UPDATED - revised illustrations to include AMSE B30.26 values for angular lifting with eyebolts
150	UPDATED - Clarified maximum included angles use for hooks and Shackles with new illustrations
151	Moved some hook use examples previously on page 150 to the bottom of this page
157	Improved upper graphic on lever hoist use to show the hoist (incorrectly) touching the obstruction
179	TYPO - Clarified that the loads shown indicate the typical load weight that a timber beam can support
212	TYPO - Updated the table on wire rope classifications (previous columns were incorrectly labeled)

Changes made in the December 2008 printing:

Page	Change/revisions
56	Revised descriptions of Tell-tails and optical fiber use
52-56	Added "Data courtesy of SLINGMAX® Rigging Solutions"
58	TYPO - changed stand to strand
63	TYPO - changed VCR to VRC

Rigging Training presented at your location!

It's easy to see why on-site training is our most requested service! We can assemble a customized training program specifically for your company, or offer several of our standard training programs:

- Basic Rigging Safety
- Rigging Fundamentals
- Advanced Rigging Workshops
- Overhead Crane Operation

With their many years of on-the-job AND classroom experience, our Training Instructors skillfully cover the training topics to provide an exceptional learning experience for your employees.

We can also offer advanced hands-on training using our portable drifting frame equipment.

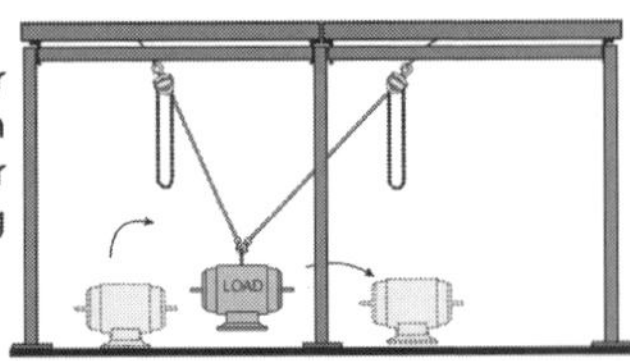
LOAD

Training programs can range from 4 hours to several days, depending on the topics and activities. We offer on-site training programs for domestic and international locations. We have classroom and workshop programs that include hands-on activities for participants to perform proper rigging practices. Since class content and your location may influence the training costs please contact us for a quote for your specific training needs.